SEISMIC TOMOGRAPHY AND MANTLE CIRCULATION

SEISMIC TOMOGRAPHY AND MANTLE CIRCULATION

PROCEEDINGS OF
A ROYAL SOCIETY DISCUSSION MEETING
HELD ON 13 AND 14 APRIL 1988

ORGANIZED BY R. K. O'NIONS, F.R.S.,
R. W. CLAYTON AND B. PARSONS
AND EDITED BY R. K. O'NIONS, F.R.S.,
AND B. PARSONS

LONDON
THE ROYAL SOCIETY
1989

Printed in Great Britain for the Royal Society
by the
University Press, Cambridge

ISBN 0 85403 382 3

First published in *Philosophical Transactions of the Royal Society of London*,
series A, volume 328 (no. 1599), pages 289–442

♾ The text paper used in this publication meets the minimum requirements of American National Standard for Information Sciences–Permanence of Paper for Printed Library Materials, ANSI Z39.48-1984.

British Library Cataloguing in Publication Data

Seismic tomography and mantle circulation.
1. Earth. Interior. Measurement. Use of seismic waves
I. O'Nions, R. K. II. Parsons, B.
551.1′1

ISBN 0-85403-382-3

Published by the Royal Society
6 Carlton House Terrace, London SW1Y 5AG

PREFACE

The pattern of convective circulation in the mantle and its relation to the thermal and chemical history of the Earth is one of the basic problems in geology. Over the past few years, seismic tomography has begun to produce intriguing maps of lateral variations in seismic velocities within the Earth. Part of these variations must be caused by horizontal temperature variations so that, although one can only be confident about the longest wavelength features at present, these new observations offer the tantalizing prospect of being able to map out the structure of mantle convection. During the next decade, a large number of new digital, broad-band seismometers will be installed worldwide, and one can expect that seismic tomography will be able to resolve increasingly smaller-scale features inside the Earth once these data become available. It seemed a good time, therefore, to review the achievements of seismic tomography to date. In particular we aimed to encourage a discussion that would enable non-seismologists to appreciate better the uncertainties in the seismic velocity structures obtained by seismic tomography, and to focus on the interesting problems that have arisen through the interaction of seismic tomography with other disciplines.

Seismic tomography has stimulated much research in other branches of the Earth sciences. On the persistent question of whether the large-scale circulation extends throughout the whole mantle, or whether flow cannot occur across the boundary between upper and lower mantle, circulation models, developed by using seismic tomography to estimate the driving density variations and with the geoid as a constraint, seem to favour mantle-wide convection. However, other evidence, such as that provided by the isotopic ratios of rare gases observed in mantle samples, is more simply explained assuming layered mantle convection. Estimates of density and temperature variations from the observed seismic velocity variations depend on material properties under the conditions of temperature and pressure found in the deep mantle, and the relation between the magnitude of the P- and S-wave velocity variations found for the lower mantle provide a direct constraint on the conditions to be found there. An unexpected connection has been made between features in the Earth's magnetic field and temperature variations near the core–mantle boundary, and topography on it, inferred from seismic tomography. The Discussion Meeting on seismic tomography and mantle circulation provided a good opportunity therefore to bring together people working in seismology, fluid dynamics, geomagnetism, mineral physics, and geochemistry. We hope that the meeting, and this volume, will have helped further the dialogue between the numerous disciplines involved in research into mantle circulation.

December 1988

R. K. O'Nions
B. Parsons

PREFACE

CONTENTS

[Three plates]

Phil. Trans. R. Soc. Lond. A **328**, 291–308 (1989)
Printed in Great Britain

Seismic modelling of the Earth's large-scale three-dimensional structure

BY J. H. WOODHOUSE AND A. M. DZIEWONSKI

Department of Earth and Planetary Sciences, Harvard University, 20 Oxford Street, Cambridge, Massachusetts 02138, *U.S.A.*

[Plates 1 and 2]

Several different kinds of seismological data, spanning more than three orders of magnitude in frequency, have been employed in the study of the Earth's large-scale three-dimensional structure. These yield different but overlapping information, which is leading to a coherent picture of the Earth's internal heterogeneity. In this article we describe several methods of seismic inversion and intercompare the resulting models.

Models of upper-mantle shear velocity based upon mantle waveforms (Woodhouse & Dziewonski (*J. geophys. Res.* **89**, 5953–5986 (1984))) ($f \lesssim 7$ mHz) and long-period body waveforms ($f \lesssim 20$ mHz; Woodhouse & Dziewonski (*Eos, Wash.* **67**, 307 (1986))) show the mid-oceanic ridges to be the major low-velocity anomalies in the uppermost mantle, together with regions in the western Pacific, characterized by back-arc volcanism. High velocities are associated with the continents, and in particular with the continental shields, extending to depths in excess of 300 km. By assuming a given ratio between density and wave velocity variations, and a given mantle viscosity structure, such models have been successful in explaining some aspects of observed plate motion in terms of thermal convection in the mantle (Forte & Peltier (*J. geophys. Res.* **92**, 3645–3679 (1987))). An important qualitative conclusion from such analysis is that the magnitude of the observed seismic anomalies is of the order expected in a convecting system having the viscosity, temperature derivatives and flow rates which characterize the mantle.

Models of the lower mantle based upon P-wave arrival times ($f \approx 1$ Hz; Dziewonski (*J. geophys. Res.* **89**, 5929–5952 (1984)); Morelli & Dziewonski (*Eos, Wash.* **67**, 311 (1986))) SH waveforms ($f \approx 20$ mHz; Woodhouse & Dziewonski (1986)) and free oscillations (Giardini *et al.* (*Nature, Lond.* **325**, 405–411 (1987); *J. geophys. Res.* **93**, 13716–13742 (1988))) ($f \approx 0.5$–5 mHz) show a very long wavelength pattern, largely contained in spherical harmonics of degree 2, which is present over a large range of depths (1000–2700 km). This anomaly has been detected in both compressional and shear wave velocities, and yields a ratio of relative perturbations in v_S and v_P in the lower mantle in the range 2–2.5. Such values, which are much larger than has sometimes been assumed, roughly correspond to the case that perturbations in shear modulus dominate those in bulk modulus. It is this anomaly that is mainly responsible for the observed low-degree geoid undulations (Hager *et al. Nature, Lond.* **313**, 541–545 (1985))). In the upper part of the lower mantle this pattern consists of a high-velocity feature skirting the subduction zones of the Pacific and extending from Indonesia to the Mediterranean, with low velocities elsewhere; thus it appears to be associated with plate convergence and subduction. The pattern of wave speeds in the lowermost mantle is such that approximately 80% of hot spots are in regions of lower than average velocities in the D″ region.

The topography of the core–mantle boundary, determined from the arrival times of reflected and transmitted waves (Morelli & Dziewonski (*Nature, Lond.* **325**, 678–683 (1987))), exhibits a pattern of depressions encircling the Pacific, having an amplitude of approximately ± 5 km, which has been shown to be consistent with the stresses induced by density anomalies inferred from tomographic models of the lower mantle (Forte & Peltier (*Tectonophysics* (In the press.) (1989))).

By using both free oscillations (Woodhouse *et al.* (*Geophys. Res. Lett.* **13**, 1549–1552 (1986))) and travel-time data (Morelli *et al.* (*Geophys. Res. Lett.* **13**, 1545–1548 (1986))), the inner core has been found to be anisotropic, exhibiting high velocities for waves propagating parallel to the Earth's rotation axis and low velocities in the equatorial plane.

Tomographic models represent an instantaneous, low-resolution image of a convecting system. They require for their detailed interpretation knowledge of mineral and rock properties that are, as yet, poorly known but that laboratory experiments can potentially determine. The fact that the present distribution of seismic anomalies must represent the current configuration of thermal and compositional heterogeneity advected by mantle flow, imposes a complex set of constraints on the possible modes of convection in the mantle of which the implications have not yet been worked out; this will require numerical modelling of convection in three dimensions, which only recently has become feasible. Thus the interpretation of the 'geographical' information from seismology in terms of geodynamical processes is a matter of considerable complexity, and we may expect that a number of the conclusions to be drawn from the seismological results lie in the future.

Introduction

Plate tectonics, which, over the past 20 years, has provided the framework for understanding large-scale geological processes, describes the motion of large, rigid sections of the Earth's crust and uppermost mantle. The plates are transported by mantle convection currents, but a definitive understanding of mantle convection has not yet been reached. Seismic tomography offers the opportunity to investigate the interior of the convective system by mapping, in three dimensions, the seismic wave velocity variations. These are related to temperature and

---

Note on plates 1 and 2. The maps shown in these plates are labelled with the following keys: *Depth*: in kilometres; *parameter*: v_P = P-velocity; v_S = S-velocity; CMB = core-mantle boundary topography; *Div* = plate velocity divergence; *data type*: T = arrival-time data; W = waveform data; M = normal mode data. In plate 2 the scale at the bottom of each panel applies to each of the plates above it. In the case of v_P the range of the scale is that given by the upper pair of labels on the scale ($\pm 0.5\%$ or $\pm 0.3\%$), and in the case of v_S the range is given by the lower pair of labels ($\pm 1.0\%$ or $\pm 0.6\%$).

Description of plate 1

Plate 1. (*a*, *b*, *c*, *d*) Three-dimensional sections of models M84C and L02.56, together with a schematic illustration of the anisotropic properties of the inner core. The depth of the section, in kilometres, is indicated. In the upper-mantle panels (550 km) the depth scale is exaggerated by a factor of 5. Plate boundaries (yellow) are indicated. (*e*) Relative S-velocity perturbations in the model U84L85/SH at the depth of 150 km, characteristic of the uppermost mantle. Plate boundaries are indicated. (*f*) The truncated spherical harmonic expansion, to degree 8, of the horizontal divergence of the instantaneous plate velocity field (Minster & Jordan 1978; Forte & Peltier 1987). Units are 10^{-9} a^{-1}. (*g*) The plate divergence field, spherical harmonic degrees 2 and 3 only. See caption to (*f*). (*h*) An upper-mantle section through the S-velocity model U84L85/SH, taken along the great circle passing horizontally through the centre of the accompanying map; depth, running vertically, is in the interval 22 km–670 km; vertical exaggeration is 20:1. (*i*) Topography of the core–mantle boundary from the study of Morelli & Dziewonski (1987*a*). Blue areas are elevated and orange areas are depressed; see scale. The model contains spherical harmonics up to degree 4. (*j*) The predicted surface divergence field (degrees 2 and 3 only) using the model U84L85/SH and the density/velocity scaling $d \ln \rho / d \ln v_S = 0.16$ in the upper mantle and 0.20 in the lower mantle. Upper-mantle viscosity is 10^{21} Pa s and lower mantle viscosity is 3×10^{22} Pa s. See caption to plate 1*f*, *g*.

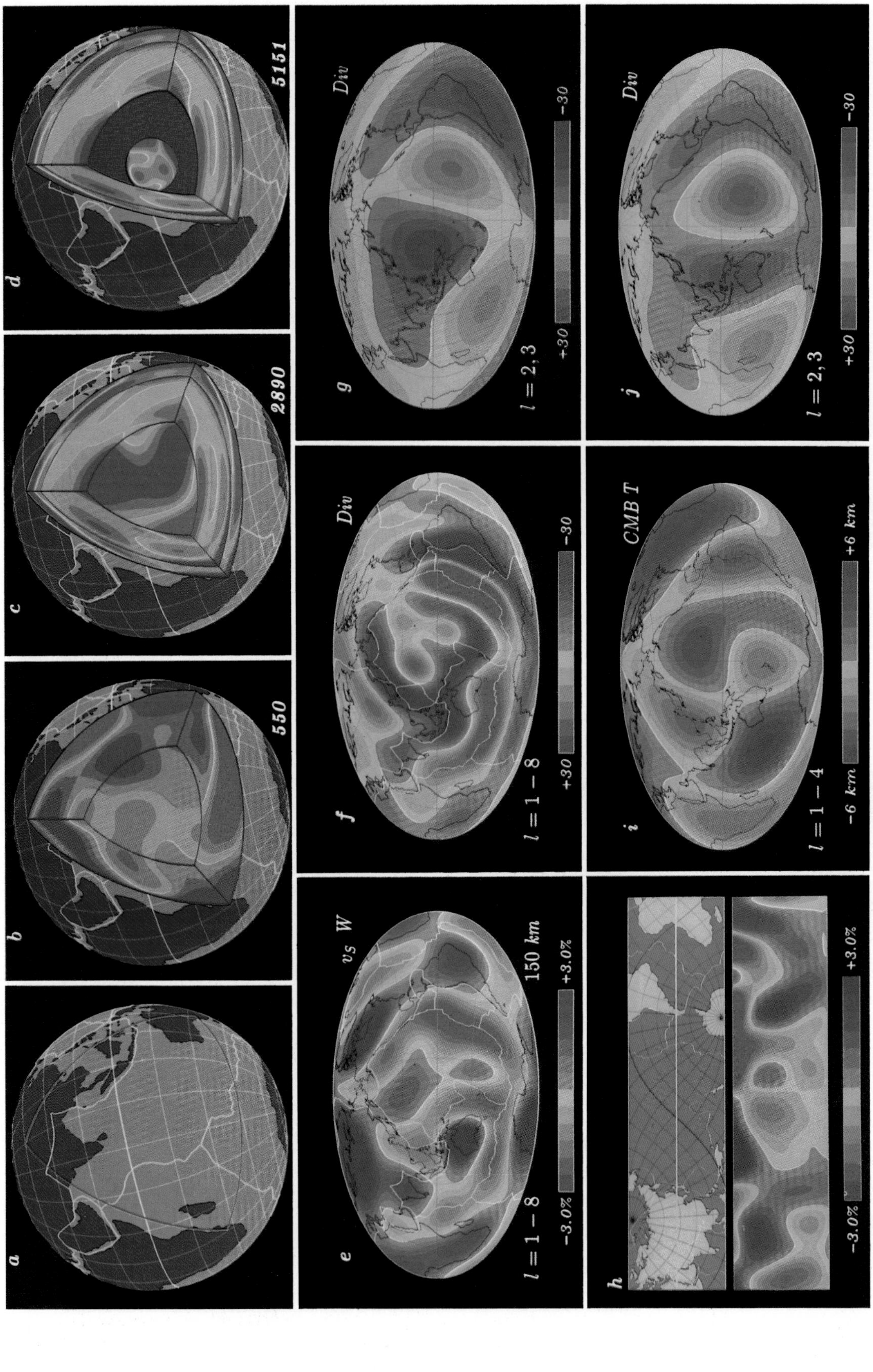

PLATE 1. For description see opposite.

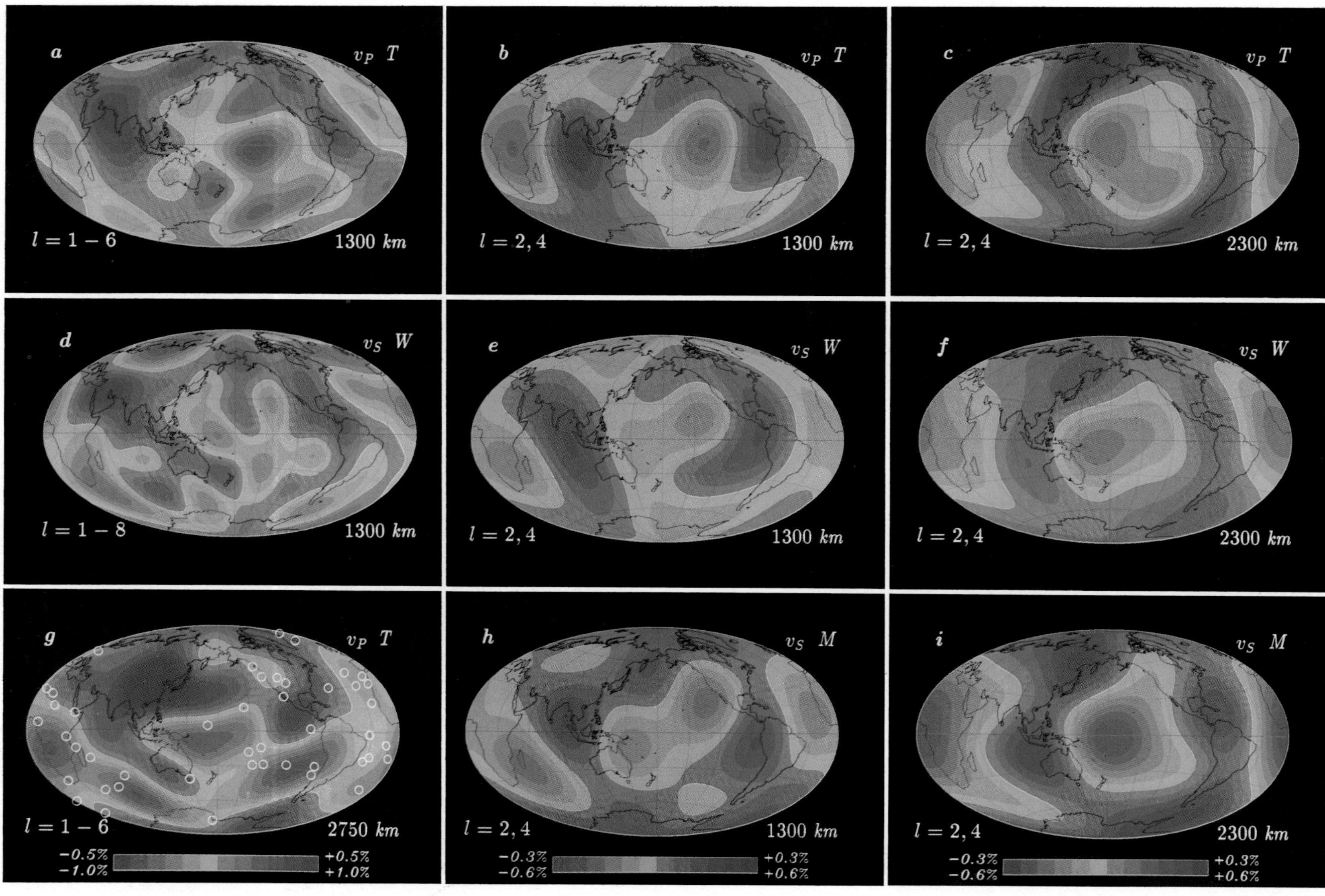

PLATE 2. For description see previous page.

composition and, in particular, to the density variations which provide the driving force of mantle convection.

Following large and intermediate earthquakes, seismic waves can be observed at stations throughout the world and significant progress has recently been made in solving the inverse problem in which very large collections of seismic data are used to reconstruct an image of the pattern of high and low wave velocities in three dimensions. Although the global images made up to now contain only the largest scale features of the Earth's three-dimensional structure, some of the conclusions to be drawn from this new 'geographical' information are now emerging. By virtue of improvements in global coverage by seismic instruments and of increasingly powerful analysis and computational techniques, there is the prospect of more detailed images in the future. In this article we discuss a number of techniques that have been applied to the global tomographic inverse problem and the models that they have lead to.

Plate 1*a*, *b*, *c*, *d* shows a composite of models of the upper-mantle S-velocity (M84C: Woodhouse & Dziewonski 1984) and the lower-mantle P-velocity (L02.56: Dziewonski 1984), and a schematic representation of the anisotropic properties of the inner core (Woodhouse *et al.* 1986; Morelli *et al.* 1986). The colour scale is such that red colours indicate lower than average wave velocities at a given depth and blue indicates higher than average velocities. These figures illustrate the scale lengths of global heterogeneity that have been resolved. In the upper mantle, the model is expanded up to degree 8 in spherical harmonics and as a cubic polynomial in depth, corresponding to nominal resolving lengths of roughly 2500 km horizontally and 150 km vertically. In the lower mantle the model is expanded up to degree 6 and as a quartic polynomial in depth, corresponding to resolving lengths of roughly 3000 km and 400 km respectively. These models represent, therefore, the result of a spatial filter applied to the true state of heterogeneity in the Earth. Such filtered versions of reality would, perhaps, be of limited interest if heterogeneity on smaller scales were dominant, but it appears that the spatial spectrum of heterogeneity contains very strong long wavelength components. In the upper mantle, for instance, S-velocity variations in spherical harmonic degrees up to 5 are as large as $\pm 4\%$ in the upper 150 km and are much larger than the expected signature of subducted slabs in this range of wavelengths. Similarly in lower-mantle P-velocity spherical harmonic degrees 2 and, to a lesser extent, 4 are dominant terms and have amplitudes of the order $\pm 0.5\%$ (plate 2*c*).

The study of lateral heterogeneity is of great significance to seismology. For example, in the investigation of earthquakes lateral heterogeneity like an imperfect lens, can distort the image of an event. Estimates of location, fault length, and the pattern of stress release can be false if the medium is inadequately known. Even the introduction of corrections for long wavelength

---

Description of plate 2

Plate 2. (*a*) The P-velocity travel-time model V3.I of Morelli & Dziewonski (1986) at depth 1300 km. The model contains spherical harmonics up to degree 6. (*b*) Model V3.I, depth 1300 km, degrees 2 and 4 only; see caption to (*a*). (*c*) Model V3.I, depth 2300 km, degrees 2 and 4 only; see caption to (*a*). (*d*) The S-velocity model U84L85/SH of Woodhouse & Dziewonski (1986) at depth 1300 km. The model contains spherical harmonics up to degree 8. (*e*) Model U84L85/SH, depth 1300 km, degrees 2 and 4 only; see caption to (*d*). (*f*) Model U84L85/SH, depth 2300 km, degrees 2 and 4 only; see caption to Plate (*d*). (*g*) Model V3.I, depth 2750 km (the base of the mantle), all degrees 1–6; hot spots are indicated; see caption to (*a*). (*h*) 'Model 1' of Giardini *et al.* (1987), S-velocity from free oscillation data, spherical harmonic degrees 2 and 4 only, depth 1300 km. (*i*) 'Model 1', depth 2300 km; see caption to (*h*).

lateral heterogeneity can result in shifts in inferred epicentres by as much as 20 km and changes in origin times of more than 1 s.

Outside the field of seismology, the recent results on the Earth's three dimensional structure have an impact on other fields of Earth sciences. Several examples of such linkage follow.

Mantle convection

Under the assumption that seismic anomalies are proportional to density perturbations, they provide constraints on the modelling of mantle convection and on the viscosity distribution in the mantle (Richards & Hager 1984; Forte & Peltier 1987, 1989; Hager & Clayton 1989).

Petrology and geochemistry

Models of seismic anomalies in the upper mantle show the oceanic ridges to be the dominant regions of low velocity at shallow depth. The long-wavelength models discussed in this article have the potential to provide integral constraints on petrological and thermal models of the ridge systems. Deep high-velocity anomalies are associated with the continental shields, apparently confirming the hypothesis of 'continental roots' (Jordan 1975, 1978*a*). About 80% of hot spots occur over regions of lower than average wave velocities near the core–mantle boundary (see plate 2*g*). There also appears to be a correlation between a band of low-velocity anomalies near the core–mantle boundary, in the latitude band from 10° S to 30° S, and the occurrence of a large-scale isotopic anomaly (Dupal anomaly: Hart 1984; Castillo 1988).

Geomagnetism

The regions at the core–mantle boundary where the magnetic field changes with time most rapidly coincide with low-velocity seismic anomalies, which, presumably, represent regions of elevated temperature in the lowermost mantle. The inference has been made that the thermal state of the lowermost mantle determines the stability of convection patterns, and hence of the geomagnetic field, in the outer core (Bloxham & Gubbins 1987).

Gravity

It was early recognized (Dziewonski *et al.* 1977) that there was a strong correlation between P-velocity anomalies in the lower mantle and the long-wavelength geoid undulations, but that the correlation had the opposite sign from that expected if high velocities correspond to high densities as would be the case if temperature fluctuations were responsible for the seismically observed anomalies. It has been shown, however (Pekeris 1935; Richards & Hager 1984; Hager *et al.* 1985), that density anomalies embedded in a fluid mantle induce deflections of the free surface and of the core–mantle boundary, which can change the sign of the inferred geoid perturbations; thus lower-mantle heterogeneity has been identified as a major cause of long-wavelength geoid anomalies.

Mineral physics

An example of an *in situ* measurement made possible through seismic tomography is the ratio of relative perturbations in the shear and compressional velocities: $\mathrm{d}\ln v_S/\mathrm{d}\ln v_P$. The tomographic results yield a ratio much higher than determined, at relatively low pressures, in the laboratory (Anderson *et al.* 1968). It appears that under the temperature and pressure conditions appropriate for the lower mantle, the shear modulus is much more sensitive to changes in temperature than the bulk modulus. Similar observations for the upper mantle have been interpreted as being associated with partial melting (Hales & Doyle 1967).

Geodesy and astronomy

From the analysis of data on the Earth's rotation, obtained by the VLBI (very long-baseline interferometry) technique, it has been determined that the flattening of the Earth's core departs from its equilibrium value by 400 ± 100 m (Gwinn *et al.* 1986). Although the seismically determined core mantle boundary topography (Morelli & Dziewonski 1987*a*; plate 1*i*) has a range of ± 6 km, the component corresponding to excess flattening is very small, and error estimates are such that the seismic results are consistent with the geodetically determined flattening.

Studies of the large-scale three-dimensional structure of the Earth have been carried out using various kinds of seismological data, spanning more than three orders of magnitude in frequency (1 Hz–0.0005 Hz). These are (i) large collections of P (and PKP, PKIKP, PcP) travel times, (ii) measurements of phase and group delays and amplitude anomalies of surface waves and measurements of the locations of spectral peaks of fundamental modes, interpreted asymptotically, (iii) complete waveforms of mantle waves, used as data in a least-squares inversion, (iv) complete waveforms of long-period body waves and (v) complete spectra of split multiplets in the Earth's free oscillation spectrum. A recent review is by Dziewonski & Woodhouse (1987).

Studies of class (i) have illuminated lower-mantle P-velocity structure (Dziewonski *et al.* 1977; Dziewonski 1984; R. W. Clayton & R. Comer, unpublished) and those of classes (ii) and (iii) have led to models of upper-mantle S-velocity (Masters *et al.* 1982; Nakanishi & Anderson 1982, 1983, 1984; Woodhouse & Dziewonski 1984; Nataf *et al.* 1984, 1986). With the addition of classes (iv) and (v) (Woodhouse & Dziewonski 1986; Tanimoto 1987; Giardini *et al.* 1987, 1988) it has become possible to constrain lower-mantle S-velocity structure and thus to obtain models of the same region of the Earth by using different classes of data. This is very valuable in that it provides a check on the various modelling techniques and also allows the comparison of heterogeneity in different structural parameters in the same region. With one kind of data alone, it is often difficult to completely rule out the possibility that systematic errors or deficiencies in coverage, which are inherent in the data, degrade or corrupt the resulting models. We find, however, that the application of different techniques is yielding a coherent picture of the Earth's global heterogeneity, reinforcing the conclusions drawn from each data set alone.

In §2 below we outline the general framework of the tomographic inverse problem and in §3 we discuss some of the techniques that have been employed and intercompare the resulting models.

2. The tomographic inverse problem

Observations of seismic disturbances contain information about the earthquakes that generate them – for instance the location and origin time of the event, the orientation of faulting and so forth – and about the structure of the Earth. The mechanics of the Earth's linear, adiabatic, elastic vibrations, which describes all seismic motions except those in the immediate vicinity of an earthquake, is characterized, in the case of an isotropic material, by the bulk modulus $\kappa(\boldsymbol{x})$, the shear modulus $\mu(\boldsymbol{x})$ and the density $\rho(\boldsymbol{x})$, where $\boldsymbol{x}$ is position. At relatively high frequencies (larger than 0.2 Hz, say) wave propagation is well described in terms of packets of energy propagating along optical rays. In the solid portions of the Earth there exist compressional waves (P-waves) having velocity $v_P = [(\kappa + \frac{4}{3}\mu)/\rho]^{\frac{1}{2}}$ and shear waves (S-waves) of velocity $v_S = [\mu/\rho]^{\frac{1}{2}}$; in the fluid outer core and the ocean μ and v_S vanish and shear waves do not exist. For an anisotropic material a number of other moduli must be introduced and the wave velocities depend upon the local direction of travel and on the polarization of the wave, but these are complications that we shall neglect in this article.

To clarify the discussion of the tomographic inverse problem we here outline a simple but general formalism that encompasses all of the specific applications to be considered later. Let d_n^s represent the nth observation relating to the sth earthquake source. Data d_n^s may represent, for example, the arrival time of a particular seismic phase at the nth receiver, or it may represent the nth sample in a large array of waveforms or spectra from a number of stations. In general, d_n^s will be functionals of observed seismograms and will be very numerous. Let f_k^s denote the necessary source parameters relating to the sth source; for arrival time data f_i^s $(i = 1, 2, 3, 4)$ will consist of the location and origin time of the event and for waveform or spectral data they will also include parameters describing the size of the event and the geometry of faulting. Given a model of the Earth $\boldsymbol{E} = [v_P(\boldsymbol{x}), v_S(\boldsymbol{x}), \rho(\boldsymbol{x})]$ we may, in principle, predict the expected values of the observations:

$$d_n^s = D_n[\boldsymbol{f}^s, \boldsymbol{E}] + \epsilon_n^s, \tag{1}$$

where D_n is a known function of $\boldsymbol{f}^s$ and functional of $\boldsymbol{E}(\boldsymbol{x})$ and ϵ_n^s is observational error. In practical applications it is necessary to represent the continuous function $\boldsymbol{E}(\boldsymbol{x})$ in terms of a finite set of parameters. We shall write

$$\boldsymbol{E}(\boldsymbol{x}) = \boldsymbol{E}_0(r) + \sum_{klm} E_{klm}\, \boldsymbol{m}_{klm}(\boldsymbol{x}), \tag{2}$$

where r is radius, $\boldsymbol{E}_0(r)$ is a spherically symmetric, reference earth model and $\boldsymbol{m}_{klm}(\boldsymbol{x})$ are a chosen set of basis functions depending upon three indices k, l, m. It is possible, for example, to partition the Earth into a three-dimensional array of cells labelled by a radial index, k, a latitude index, l and a longitude index m and to define $\boldsymbol{m}_{klm}(\boldsymbol{x})$ to be $[1, 0, 0]$ in the (k, l, m)th cell and to vanish elsewhere; in this case the expansion coefficient E_{klm} would represent the deviation of P-velocity in the (k, l, m)th cell from that of the reference model. This approach has been used in both regional and global studies (see, for example, Aki *et al.* 1977; Dziewonski *et al.* 1977). Another approach, which we have employed in the studies described below, is to choose $\boldsymbol{m}_{klm}(x)$ to consist of certain smooth functions, e.g.

$$\boldsymbol{m}_{klm}(\boldsymbol{x}) = \boldsymbol{m}_k(r)\, Y_l^m(\theta, \phi), \tag{3}$$

where (r, θ, ϕ) are spherical coordinates (radius, colatitude, longitude), $Y_l^m(\theta, \phi)$ are spherical harmonics and where $\boldsymbol{m}_k(r)$ $(k = 1, 2, \ldots, K)$ are a set of (orthogonal) radial vector functions. In this case, E_{klm} represent the expansion coefficients in a spherical harmonic series, also expanded in radius.

By using any expansion of the form (2), equation (1) may be written

$$d_n^s = D_n[\boldsymbol{f}^s, \boldsymbol{E}_0 + \sum_{klm} E_{klm}\, \boldsymbol{m}_{klm}(\boldsymbol{x})] + \epsilon_n^s. \tag{4}$$

The tomographic inverse problem may be thought of as that of characterizing the set of solutions (f_i^s, E_{klm}) consistent with assumed statistical properties of the errors and with statistical *prior information* on the model parameters (see Tarantola & Valette 1982*a*). If the corresponding distributions are assumed to be gaussian then (4) leads to a version of the least-squares technique. If the covariances $\mathrm{cov}(\epsilon_n^s, \epsilon_{n'}^{s'})$ are specified it is, in principle, straightforward to linearly transform the data in such a way that the transformed errors are independent gaussian variates of uniform variance σ^2, say. Similarly a linear transformation of the set of unknowns (f_i^s, E_{klm}) may be found such that the prior information consists of requiring that the transformed unknowns be independent samples of a gaussian distribution of variance η^2, say. In the interests of being concise we shall here assume that the data and model parameters in (4) already possess these simple statistical properties and that there is no prior information on source parameters. In this case the maximum likelihood estimate of the unknowns (f_i^s, E_{klm}), the expectation of the *a posteriori* distribution, is that which minimizes

$$L \equiv \sigma^{-2} \sum_{ns} \{d_n^s - D_n[\boldsymbol{f}^s, \boldsymbol{E}_0 + \sum_{klm} E_{klm} \boldsymbol{m}_{klm}(\boldsymbol{x})]\}^2 + \eta^{-2} \sum_{klm} \{E_{klm} - E_{klm}^0\}^2, \tag{5}$$

where E_{klm}^0 is a prior heterogeneous model (the expectation of the *a priori* distribution of model parameters, usually zero in applications). This has the form

$$L = \sigma^{-2} |\boldsymbol{d} - \boldsymbol{D}(\boldsymbol{x}, \boldsymbol{y})|^2 + \eta^{-2} |\boldsymbol{x} - \boldsymbol{x}^0|^2, \tag{6}$$

where the algebraic vectors $\boldsymbol{d}, \boldsymbol{x}, \boldsymbol{y}$ correspond to the data, the model parameters and the source parameters for all sources, respectively.

Because, in general, $\boldsymbol{D}(\boldsymbol{x}, \boldsymbol{y})$ is a nonlinear function of its arguments, the minimization of (6) is an iterative process (Tarantola & Valette 1982*a*, *b*) in which the partial derivatives of $\boldsymbol{D}(\boldsymbol{x}, \boldsymbol{y})$ are required. For travel-time data these are provided by Fermat's principle, by which the travel-time increment corresponding to a perturbation in wave speed may be calculated in terms of an integral along the unperturbed ray path; in general, the partial derivatives must be obtained through a theoretical analysis of the wave propagation problem and will be given by a form of the Born approximation. It is of interest to note that whereas the accuracy of these derivatives will influence the rate of convergence to a minimum, it will not affect the value $(\boldsymbol{x}_\infty, \boldsymbol{y}_\infty)$ say, to which, let us assume, the iterates converge. Assuming convergence, the accuracy of the solution minimizing (6) depends only upon that with which it is possible to calculate the solution of the forward problem, $\boldsymbol{D}(\boldsymbol{x}, \boldsymbol{y})$.

This formalism also allows the estimation of the *a posteriori* statistics of $(\boldsymbol{x}, \boldsymbol{y})$ provided that it is sufficient to represent $\boldsymbol{D}(\boldsymbol{x}, \boldsymbol{y})$ by a linear approximation in the neighbourhood of $(\boldsymbol{x}_\infty, \boldsymbol{y}_\infty)$.

In the event that no prior information or insufficient prior information is available, the solution minimizing (6) may be indeterminate. In this case an important concept is that of *resolution* (Backus & Gilbert 1968; Jackson 1979; Gubbins & Bloxham 1985). Recognizing that the minimization of (6) does not yield a useful estimate of $\boldsymbol{x}$, one seeks a transformation

$$\boldsymbol{x}' = \boldsymbol{R}\boldsymbol{x} \tag{7}$$

such that $\boldsymbol{x}'$ has suitably small *a posteriori* errors. The operator $\boldsymbol{R}$, which ideally is close to the identity, represents linear combinations of model parameters which are estimable within a chosen level of error. This operator is commonly chosen to coincide with that which results from the incorporation of fictitious *prior information*, although it is possible to define other strategies for choosing $\boldsymbol{R}$ to achieve desired resolution and error characteristics in the solution. The quality of a solution $\boldsymbol{x}'$ is then characterized by its errors (more generally by its covariance matrix) and its resolution operator $\boldsymbol{R}$. The concept of resolution does not easily generalize to nonlinear problems, and thus one needs to perform such analysis in the neighbourhood of $(\boldsymbol{x}_\infty, \boldsymbol{y}_\infty)$ using a linearized approximation to $\boldsymbol{D}(\boldsymbol{x}, \boldsymbol{y})$; in this case the results of such calculations must be interpreted with caution, because they do not correctly represent the statistical estimation problem.

The usefulness of the formalism outlined here is limited by our inability, in most cases, to realistically characterize the statistical properties of the errors ϵ_n^s. Far from being independent gaussian samples there are invariably many potential sources of systematic error which are essentially impossible to discover or to quantify. An example would be the potential biasing effect on tomographic models caused by the fact that seismic sources preferentially occur in or near subduction zones, the locations of which are highly correlated with larger-scale patterns of mantle heterogeneity.

The 'errors' in (1) also include components that are not properly described as observational errors. In the case that complete seismograms are used as data, the function D_n in (1) consists of an algorithm for the calculation of synthetic seismograms, and although at long periods it is possible, in principle, to calculate accurate synthetic seismograms, such calculations are too laborious to be feasible in application to the inversion of large data sets. Thus applications of this technique have used asymptotic (ray theoretic) approximations to the effects of heterogeneity which undoubtedly are often subject to appreciable theoretical errors.

Probably the largest source of discrepancy between data and model predictions arises from unmodelled structure, which is not describable in terms of the chosen set of basis functions $\boldsymbol{m}_{klm}(\boldsymbol{x})$. An example is the potential effects due to anisotropy on a model that does not contain anisotropic parameters. The answer to such difficulties is not always to include more model parameters, because in this case many of them may become impossible to usefully constrain. In a linear problem it is possible to calculate which linear combinations of model parameters can be estimated with a given level of error, but the usefulness of this is questionable; new scientific conclusions are unlikely to result from knowing the values of a number of linear combinations of disparate model parameters, which are, however, insufficient to give estimates of any individual one. The information contained in a given data set, if all potential model parameters are included, consists of a statistical distribution in a model space of high dimension; but such a distribution is impossible to visualize and thus it is impossible to make any useful inference.

Instead, in our view, one must proceed on the basis of certain hypotheses: for example, the hypothesis that errors in synthetic seismogram techniques do not vitiate inversion results, or the hypothesis that the effects of mantle anisotropy do not corrupt the results of an inversion for isotropic structure. There arise opportunities to test such hypotheses when it proves to be possible to investigate the same region of the Earth using different kinds of data and different modelling techniques. For example, a very-long-wavelength, high-velocity structure in the lower mantle, bearing a clear relationship to the locations of subduction zones (see below), has been independently found using P-wave arrival times, vertically polarized long-period P–SV waveforms, horizontally polarized SH waveforms and spectra of free oscillations. These data sets are fundamentally different in their sensitivities to unmodelled effects, and different theoretical techniques are used in each case. Low-degree free oscillations, for example, are insensitive to the locations of sources; vertically and horizontally polarized shear waves have greatly different sensitivities to anisotropy and different from that of P-waves; the waveform data consist largely of multiple reflections of S-waves by the free surface and therefore their sensitivity to mantle structure differs greatly from that of the direct P-wave; in addition they are much less sensitive to near-surface effects. By the intercomparison of models built by using different techniques it is possible to test the validity of the approximations or simplifying assumptions which, out of necessity, have been made.

The above statements of caution regarding the estimation of errors and resolution are not peculiar to the tomographic problem. In the determination of the hypocentral location of earthquakes, for example, formal error analysis invariably underestimate true uncertainties, for the same reasons as discussed above. Even if the Earth were perfectly spherically symmetric, the fact that measures of uncertainty have not been determined for spherically symmetric models of the Earth would make it impossible to correctly quantify the uncertainty in hypocentral locations. These problems are greatly compounded by the presence of lateral heterogeneity and anisotropy. We discuss this issue in the context of tomographic modelling because in the studies discussed here the models are formally overdetermined to such an extent that the conventional error estimates, scaling inversely as the square root of the number of data, give unrealistically small estimates of uncertainty. What is usually meant by an error estimate is actually a measure of the sensitivity of the solution to statistical fluctuations in the data. In the tomographic problems outlined in this article this is small, and is only a minor component of the true uncertainty as measured, for example, by independent inversions from different partitions of a given data set (see, for example, Woodhouse & Dziewonski 1984) or by the intercomparison of models constructed with different kinds of data. It is our aim in this discussion simply to note that in attempting to quantify the uncertainties in tomographic models, conventional formulae based upon the above statistical formalism, are usually inadequate, owing to the failure of the 'errors' in the data to satisfy the underlying statistical assumptions.

3. Tomographic techniques and models

The foregoing analysis identifies aspects that the particular techniques, to be discussed here, have in common. A technique is characterized by specifying (i) the data set, (ii) the theoretical algorithm $D_n[\boldsymbol{f}^s, \boldsymbol{E}]$, (iii) the basis functions $\boldsymbol{m}_{klm}(\boldsymbol{x})$ and (iv) the statistical assumptions concerning the distribution of 'errors' and the choice of resolution operator. Naturally, if modelling results are greatly sensitive to the somewhat arbitrary choices that enter into this formulation, the results are called into question. Here we discuss several specific techniques and present some comparisons of corresponding models.

3.1. *Arrival times*

Arrival times of the principal seismic phases are read by station operators around the world and are reported to the International Seismological Centre (ISC), which publishes them both in its printed bulletin and in computer-readable form. These voluminous catalogues, containing readings from more than 1000 stations, have proved to be a valuable resource in determining both the spherical and the aspherical structure of the Earth. The seismic phases from which such readings are made have a characteristic period of approximately 1 s.

Analysis of the arrival times of the direct P-wave have led to models of lower-mantle P-velocity (Dziewonski *et al.* 1977; Dziewonski 1984; Morelli & Dziewonski 1986). The data set assembled by Morelli & Dziewonski (1986), for example, consists of approximately 1.7×10^6 selected readings, relating to some 26000 events. In selecting and reducing the data, steps are taken to desensitize the solution to the potential effects of crustal structure beneath the stations and to the effects of subducted slabs (see Dziewonski 1984; Morelli & Dziewonski 1987*b*). The theoretical technique used in this problem is that given by geometrical ray theory, linearized by means of Fermat's principle, and the basis functions correspond to a spherical harmonic expansion of P-velocity up to degree 6 and a radial representation as a quartic polynomial in the depth range corresponding to the lower mantle.

Some of the properties of the model V3.I of Morelli & Dziewonski (1986), which are shared by the similar model L02.56 of Dziewonski (1984), are illustrated in plate 2*a*, *b*, *c*, *g*. Plates 2*a* and 2*g* are maps at the depths 1300 km and 2700 km, which are representative of the top and the bottom of the lower mantle, respectively. Plates 2*b* and 2*c* show the components of the model in spherical harmonic degrees 2 and 4 at depths 1300 and 2300 km. The pattern of high velocities encircling the Pacific, largely contained in degrees 2 and 4, is a dominant feature of such models and the pattern of plate 2*c* (2300 km depth) is relatively constant over a large range of depths, changing appreciably only within a few hundred kilometres of the top and the bottom of the lower mantle. Plate 2*g* shows the distribution of hot spots superimposed on the model at the base of the lower mantle, suggesting an association between the locations of hot spots with regions of low P-velocity in the lowermost mantle.

3.2. *Seismic waveforms*

Starting in the mid-1970s, the Global Digital Seismograph Network (GDSN), operated by the United States Geological Survey, and the International Deployment of Accelerometers (IDA) network, operated by the University of California, San Diego, have provided continuous, digital, long-period seismic data from a global array of receivers. During the 1980s the French GEOSCOPE network has furnished additional three-component long-period data. The digital recordings from these networks have made it possible, for the first time, to assemble very large waveform data bases for use in seismic inversion.

Theoretical studies of the free oscillations of an aspherical Earth (see Dahlen 1968) and the development of an asymptotic formalism for the interpretation of frequency shifts due to heterogeneity (Jordan 1978*b*; Silver & Jordan 1981) led to the discovery of a strong degree 2 pattern in the phase-velocity variations of fundamental-mode Rayleigh waves having periods longer than about 220 s (Masters *et al.* 1982), which was identified as originating in the transition zone (400–650 km depth).

Employing somewhat higher frequencies and using a technique capable of resolving odd-degree heterogeneity as well as even degrees and of incorporating overtone data in addition to the fundamental mode branch, Woodhouse & Dziewonski (1984) constructed a three-dimensional model of upper-mantle S-velocity containing degrees up to 8 and represented as a cubic polynomial in depth. The technique developed for this study is based upon least-squares fitting of entire waveforms of very long period ($f < \frac{1}{135}$ Hz) 'mantle waves', world-circling Rayleigh and Love waves sensitive principally to upper-mantle structure. An example of such data, together with synthetic seismograms for three models, is shown in figure 1. The principal information in this kind of data is contained in mismatches in phase between data and

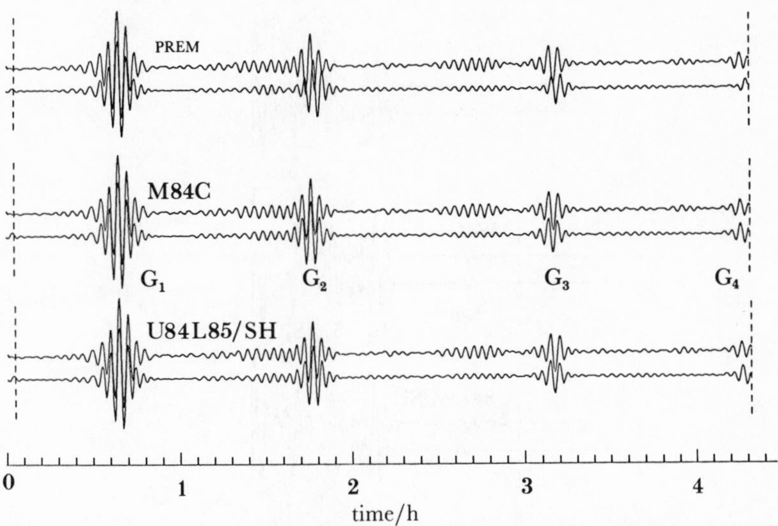

FIGURE 1. An example of very-long-period 'mantle wave' data used in the waveform inversion study discussed in the text. The data are from the station NWAO (Western Australia) of the Global Digital Seismographic Network, following an event on the Mid-Atlantic Ridge on 3 June 1981; epicentral distance is 99.5°. In each pair of traces the top trace is the data, low-pass filtered by using a cut-off frequency $\frac{1}{135}$ Hz and the bottom trace is a theoretical seismogram for a particular Earth model. The two observed horizontal components of motion have been combined to yield the transverse component. The large arrivals are the world-circling Love waves G_1, G_2 etc. In the top pair of traces the synthetic seismogram is calculated for the spherically symmetric model PREM of Dziewonski & Anderson (1981); large phase errors are evident in the major group arrivals. In the middle pair of traces (model M84C; Woodhouse & Dziewonski 1984) these phase misalignments have been largely eliminated, and similarly in the bottom pair, which corresponds to the model based upon SH waves discussed in the text. Note that the model U84L85/SH matches wave data approximately to the same degree as the model M84C, which is based upon mantle wave data alone; U84L85/SH also includes information from body waves such as those shown in figure 2.

theoretical seismograms, which are representative of certain averages along the source–receiver path and with depth of the three-dimensional variations in v_S (mainly) and v_P. The modelling strategy succeeds in greatly reducing such discrepancies, for a large suite of data, by adjusting the three dimensional distribution of variations in v_S.

More recently another kind of data consisting of long-period body waves ($f \leqslant \frac{1}{45}$ Hz) together with a larger set of mantle wave data have been used to construct global models of v_S for both the upper and the lower mantle (Woodhouse & Dziewonski 1986). Examples of the body-wave data are shown in figure 2. The largest arrivals in such data are multiple reflections of S-waves from the free surface. Woodhouse & Dziewonski (1986; the definitive versions of these models are in preparation) have used a data set consisting of some 6600 mantle wave records and 8500 body wave records from 220 globally distributed events. Analysis has been performed separately on the horizontally polarized (SH) and vertically polarized (P–SV) subsets of the data, to assess the potential effects of anisotropy; here we shall discuss only the results for SH; the SH model is designated U84L85/SH.

The theoretical technique used in these studies is based upon an asymptotic approximation to the effects of heterogeneity (Woodhouse & Dziewonski 1984; Mochizuki 1986; Romanowicz 1987), which has as one of its advantages the property that a theoretical seismogram corresponding to a given source–receiver pair depends only upon the horizontally averaged structure along the great circle path between the source and receiver. It is this property that makes feasible the calculation of large numbers of partial derivative seismograms because, for

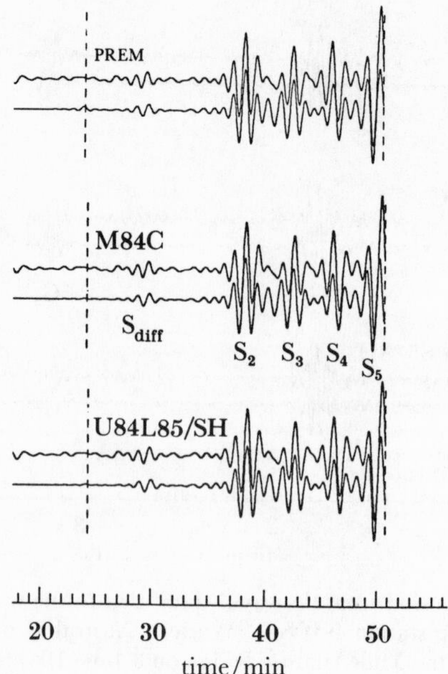

FIGURE 2. An example of long-period body-wave data; the data are the transverse component from the station KONO (Norway) following an event in the Western Pacific on 15 February 1982; epicentral distance is 123.1°. As in figure 1 the observed seismogram is the top trace of each pair, in this case low-pass filtered with a cut-off frequency $\frac{1}{45}$ Hz. The phase labelled S_{diff} is the S-wave diffracted by the core; S_n ($n = 2,3,4,5$) are the phases SS, SSS, etc., which have suffered $n-1$ reflections from the free surface between the source and the receiver. Phase errors evident in the top two pairs are greatly reduced by the model U84L85/SH, discussed in the text. See caption to figure 1.

a given source–receiver pair, it is only those linear combinations of the model parameters corresponding to the path-averaged structure that enter into the calculation. The accuracy of this approximation for fundamental modes has been investigated by Park (1987) who finds that it represents well the largest effects of heterogeneity. Further tests of this approximation and the development of more accurate techniques are certainly desirable.

The model derived by this approach from SH data is shown in plates 1*e*, *h* and 2*d*. In common with earlier models, large low-velocity anomalies are associated with mid-oceanic ridges, which, moreover, are strongest for the most rapidly spreading ridges (East-Pacific Rise, Southeast Indian Rise) and weaker for less active ridges (Mid-Atlantic Ridge). This is evident in plate 1*e*, *f* in which the shear velocity distribution at 150 km depth – characteristic of the uppermost mantle – is compared with the truncated spherical harmonic expansion of the horizontal divergence of the plate velocity field (Minster & Jordan 1978). Low-velocity anomalies are also associated with the back-arc regions of the west Pacific. In addition there are large deep high-velocity anomalies beneath the continents (Jordan 1975). Notable in their apparent absence are the traces of subduction zones; although large localized anomalies are known to be associated with subducted slabs they do not have a large signature when filtered to wavelengths greater than 2000 km, which are the horizontal scale lengths of the model here described. If a slab is represented by a 10% velocity anomaly 100 km wide, its filtered representation at a wavelength of 2000 km will correspond only to a 0.5% anomaly. The anomalies corresponding to ridges and continents, on the other hand, are several percent in amplitude and thus are large enough to mask the signal corresponding to the slabs. Plate 1*h* shows a section of the upper-mantle model taken along the great circle indicated in the map;

this displays examples of the deep extensions of the continental shields and of the occurrence of low-velocity anomalies associated with ridges.

In the lower mantle this model of S-velocity is highly correlated with P-velocity models based upon travel-time data, and also with models of v_P and v_S inferred from spectral splitting of free oscillations, discussed below; the three data sets, however, require an unexpectedly large ratio of v_P and v_S anomalies, $d \ln v_S / d \ln v_P \approx 2$–$2.5$, which corresponds roughly to the case in which variations in shear modulus dominate those in bulk modulus. A thermodynamic explanation of this has been suggested by Anderson (1987). Three comparisons of the P-velocity model V3.I with the S-velocity model U84L85/SH are shown in plates 2*a*, *d*, 2*b*, *e* and 2*c*, *f*; note that the contour intervals and the range of the colour scales for relative perturbations in v_S are twice those for v_P.

The top of the lower mantle reveals high-velocity anomalies following the subduction zones of the present and the geologically recent past (see plate 2*a*, *d*). These high-velocity features include the rim of the Pacific and a region stretching roughly from Indonesia to the Mediterranean that marks the Tethys convergence zone. In plates 2*a*, *d* and 1*f*, independent models of SH-velocity and P-velocity inferred from travel times may be compared with the present-day plate divergence field. It will be noted that the anomalies are displaced outwards from the Pacific relative to the current loci of subduction, and that regions such as North America and southern Eurasia, which have been stronger convergence zones in the past than they are at present, are strongly represented in the models. In North and Central America these models are in good agreement with the regional model of Grand (1987), which shows, with somewhat higher resolution, a high-velocity, N–S trending anomaly in the depth range 800–1200 km. At its southern end, in the Carribbean, this coincides with an anomaly identified by Jordan & Lynn (1974) and by Lay (1983). Jordan & Lynn (1974) and Grand (1987) have suggested that the anomaly represents material subducted beneath the North American continent, and the fact that it forms a part of a global feature skirting most of the subduction zones of the Pacific (plate 2*a*, *d*) tends to substantiate this.

This pattern blends, with increasing depth, into long-wavelength pattern of high velocities around the Pacific that continues to the core–mantle boundary. Plate 2*c*, *f* shows the degree 2 and 4 components of this pattern at the depth 2300 km.

3.3. *Free oscillations*

Individual arrival times used in constructing tomographic models are sensitive to very-short-wavelength features of Earth structure, because typical wavelengths of 1 Hz signals are less than 10 km. To determine large-scale structure one has to rely upon an inversion algorithm to correctly evaluate the appropriate averages. Similarly the typical wavelengths of the mantle wave and long-period body wave data used in waveform inversion are significantly less than the scale lengths of the resulting models. The very low-frequency signals of free oscillations, however, having wavelengths comparable to the Earth's radius, provide the opportunity to make use of information that is directly sensitive only to the longest wavelength features of Earth structure. An isolated mode of angular order 2, for example, is sensitive only to degrees 2 and 4 in the spherical harmonic expansion of the Earth's heterogeneity. This kind of data is also unique in the sense that source and receiver bias may be essentially ruled out as a potential source of systematic error.

For modes that are 'isolated' in the spectrum, i.e. that do not exchange energy with other modes, the theory of degenerate splitting (Dahlen 1968; Woodhouse & Dahlen 1978; Woodhouse & Girnius 1982) may be used to calculate the time series, and hence the spectrum, of the model multiplet in an aspherical Earth model; no ray theoretic approximations are needed. The splitting due to heterogeneity is quantified in terms of the *splitting function* (Woodhouse & Giardini 1985; Giardini *et al.* 1987, 1988; Woodhouse *et al.* 1986), which is analogous to a phase velocity distribution for the mode. A similar approach has been employed by Ritzwoller *et al.* (1986, 1988).

Figure 3 shows an example of observed and theoretical spectra used in such a study (Giardini *et al.* 1988). In each frame the phase spectrum in the interval $[-\pi, \pi]$ is shown in the top panel, the amplitude spectrum in the middle panel; broken lines are theoretical spectra for a given Earth model and solid lines are observed spectra. This particular spectral window (1.21–1.27 mHz) contains the two modal multiplets ${}_0S_7$, with 15 singlets, and ${}_2S_3$ with 7 singlets†. The relative amplitudes and spectral locations of the singlets contributing to the theoretical spectra are indicated in the bottom panel of each frame. The top pair of frames (figure 3*a*, *b*) show data from two different stations and theoretical spectra for a model which is elliptical and is rotating

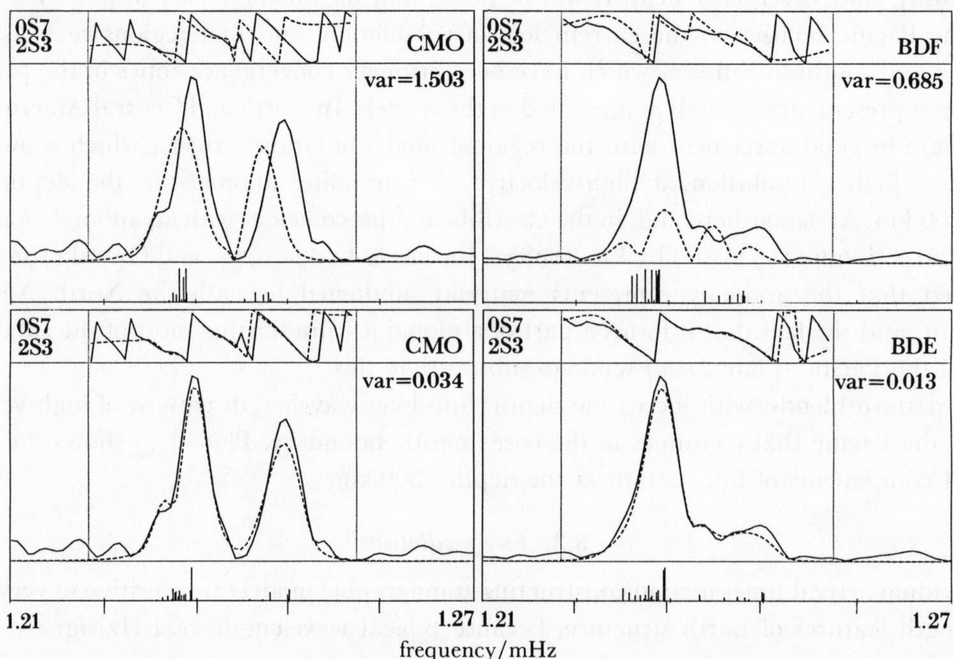

FIGURE 3. Examples of free oscillation amplitude and phase spectra from the IDA stations CMO (Alaska) and BDF (Brazil) following an event in Sumbawa, Indonesia on 19 August 1977. The top two frames show the comparison of data (solid line) with theoretical spectra (broken line) for a spherically symmetrical model, including the effects of rotation and ellipticity only. The bottom two frames show the comparison after inversion for the *splitting function* using a total of 59 spectra relating to the same pair of multiplets. The values 'var' are measures of misfit (squared misfit/squared data) in the complex spectra domain. Further details are given in the text.

† The theory of modal splitting is analogous to that describing a spherically symmetric atom subject an aspherical perturbing influence, such as a magnetic field. A degenerate multiplet ${}_nS_l$ of angular order l (c.f. total angular momentum) consists of $2l+1$ singlets belonging to the same eigenfrequency (c.f. energy level) and labelled by their azimuthal order numbers $-l, -l+1, \ldots, l-1, l$ (c.f. z-components of angular momentum). An aspherical pertubation leads to $2l+1$ eigenfunctions belonging to distinct eigenfrequencies, which, to zeroth order, are various linear combinations of the $2l+1$ degenerate eigenfunctions.

but is otherwise spherically symmetric. It will be noted that the differences between data and synthetics are very large in both amplitude and in phase. Observed spectra do not allow one to resolve individual singlets owing to the effects of attenuation and finite record length; the observed peaks are broad and represent the combined effects of the interference of the contributing singlets. The lower frames (figure 3*c*, *d*) show the same data, together with theoretical spectra after inversion for the *splitting functions* of these two multiplets. It should be noted that the data shown are but two records from a total of 59 records, from 10 events, which were used in this inversion. Thus our inability to resolve the fine structure in the spectra of individual records is compensated by the requirement that all available data in this spectral window, corresponding to different source–station pairs, must be explained simultaneously by the same distribution of singlets, having excitation amplitudes consistent with independently determined source parameters. The problem is further constrained by requiring the splitting to be due only to heterogeneity of degrees 2 and 4. Thus the problem is, at least formally, greatly overdetermined; the variance ratios (squared misfit/squared data) before and after inversion for all 59 traces are 0.689 and 0.134, respectively, in this case.

Modes sensitive primarily to mantle S-velocity structure have been used to obtain models of the lower mantle (Giardini *et al.* 1987), albeit of limited resolution (spherical harmonic degrees 2 and 4); these show strong correlation with travel-time models of v_P and, when compared quantitatively with these models, give a high value of $\mathrm{d}\ln v_S/\mathrm{d}\ln v_P \approx 2$–$2.5$, in agreement with conclusions drawn from the waveform models discussed above. Some mantle models also have non-negligible sensitivity to v_P, and provide evidence that the level of v_P heterogeneity is the same at modal periods (300–2100 s) as it is found to be at 1 s period in travel-time studies (Giardini *et al.* 1987, 1988). Modal models of v_S, at depths 1300 km and 2300 km are shown in plate 2*h*, *i* (Giardini *et al.* 1987). The two right-hand panels in plate 2 show three independent determinations of lower-mantle heterogeneity of degrees 2 and 4. The pattern of high velocities around the Pacific, largely contained in degree 2, has thus been independently obtained from three different kinds of data.

4. Discussion

From studies such as those discussed above, some of the major features of the Earth's three-dimensional structure are becoming clear and the process of interpretation is underway. In the upper mantle the continental shields are characterized by deep (*ca.* 300 km) high-velocity anomalies (see Jordan 1978*a*; Woodhouse & Dziewonski 1984) and the strongest low-velocity features are associated with the fast-spreading oceanic ridges (plate 1*e*, *h*, *g*). Whereas the latter can be understood in terms of the upwelling of hot material as the plates move apart, the deep extensions of the continents present some difficulty because it may be expected that vigorous convection would rapidly destroy them. Probably they are compositionally different, cooler and of higher viscosity than the sub-oceanic mantle (Jordan 1988). By translating the observed velocity anomalies into density anomalies and performing a dynamical calculation of fluid flow in a mantle of assumed viscosity structure, these models have been used to infer the expected fluid motion at the surface (Forte & Peltier 1987). Comparison with the poloidal component of the observed motion of the plates shows fairly good agreement at very low degree (plate 1*g*, *j*), and highly significant correlations at higher degrees and thus it has become possible to begin to understand the internal forces that drive plate motion. A limiting factor, however, is the unknown relation between density and velocity variations beneath the continents; conventional assumptions lead to difficulty in explaining surface topography and the geoid.

Much interest in geodynamics centres on the trajectory in the mantle of material subducted at convergent plate boundaries. This is because it bears on the question of how well the mantle is mixed, which is of great importance in understanding the Earth's chemical evolution. Although the debate on this issue will doubtless continue for some time to come, the most natural inference from both global (Dziewonski 1984; Woodhouse & Diezwonski 1986) and regional (see Creager & Jordan 1984, 1986; Grand 1987) seismological studies is that cold subducted material penetrates the boundary, although a significant increase in viscosity may act to reduce the velocity and to increase the cross-sectional area of the descending flow (Hager *et al.* 1985). Present day subduction takes place principally around the rim of the Pacific; models of the lower mantle show that a very-long-wavelength pattern of high velocities around the Pacific (see plate 2) persists throughout the lower mantle (Dziewonski *et al.* 1977; Dziewonski 1984; Woodhouse & Dziewonski 1986; Morelli & Dziewonski 1986; Giardini *et al.* 1987) and, moreover, that its morphology bears the closest resemblence to what may be expected due to subduction in the upper part of the lower mantle (plate 2*a*, *d*). This feature, which fills a major fraction of the Earth's volume, is undoubtedly of fundamental significance in understanding the Earth's long-term dynamics; it remains to be determined, through geodynamical modelling, whether this feature *explains* or whether it is *explained by* the fact that subduction zones have tended to be in fixed locations for long periods of Earth history. That is: is the tomographically observed anomaly to be attributed to material subducted in the same locations for long periods of time, or are the locations of convergent plate boundaries governed by a long-lived thermal anomaly in the lower mantle? Just outside the core the pattern of high and low velocities is such that approximately 80% of hot spots at the surface are above regions of lower than average velocity in the lowermost mantle, lending support to the hypothesis that hot spots are the surface manifestation of plumes rooted in the deepest part of the mantle (plate 2*g*). Instantaneous fluid dynamical calculations have had major success in explaining the very long wavelength geoid perturbations as the result of density anomalies, proportional to the seismic velocity anomalies in the lower mantle, acting as internal loads in a viscous mantle and have led to the conclusion that the viscosity of the lower mantle is substantially higher than that of the upper mantle (Hager *et al.* 1985).

The shape of the core–mantle boundary, in spherical harmonic degrees up to 4, has been determined by using the travel times of both reflected and transmitted waves (Morelli & Dziewonski 1987*a*), yielding consistent results where coverage is adequate (plate 1*i*). Fluid dynamical calculations have again been successful in explaining some aspects of the seismically inferred boundary topography (Forte & Peltier 1989). However, studies incorporating rays that have high angles of incidence (almost grazing) have shown a different pattern (Creager & Jordan 1987). This apparent discrepancy is probably a further manifestation of the complex heterogeneity in the lowermost mantle and, possibly, in the outermost core. It is important to recognize that a deflection in the core–mantle boundary produces an anomaly in the travel time of a reflected phase (PcP) that has the opposite sign from the travel-time anomaly of the transmitted wave (PKP) which intersects the boundary at the same point. A velocity anomaly in the mantle, on the other hand, would produce travel-time anomalies of the same sign in PcP and PKP. In the study of Morelli & Dziewonski (1987*a*), the demonstration that PcP gives a result similar to that obtained from PKP (and, in particular, a result of the same sign) is strong evidence that it is truly the signal of a boundary deflection, rather than that of mantle heterogeneity, that has been detected.

Although no well-documented heterogeneity has been reported in the fluid outer core – and, indeed, is not to be expected because of the inability of an inviscid fluid to maintain such heterogeneity – the inner core has been found to be characterized by an anisotropic, crystalline structure (Woodhouse *et al.* 1986; Morelli *et al.* 1986; Shearer *et al.* 1988) in which there is preferential alignment of the high velocity axes parallel to the Earth's rotation axis. This enigmatic observation is possibly evidence of low-degree convection in the inner core (Jeanloz & Wenk 1988).

We are grateful to our collaborators in much of the work reviewed in this article: D. Giardini, X.-D. Li and A. Morelli. We also thank the staff of Albuquerque Seismological Laboratory, United States Geological Survey; of the IDA Project, University of California, San Diego; and of the International Seismological Centre. These dedicated laboratories have provided the data used in the studies reported upon here. The work has been supported by the following grants from the United States National Science Foundation: EAR83-17594, EAR86-18829, EAR87-08622, EAR87-21301.

References

Aki, K., Christofferson, A. & Husebye, E. S. 1977 *J. geophys. Res.* **82**, 277–296.
Anderson, D. L. 1987 *Physics Earth planet. Inter.* **45**, 307–323.
Anderson, O. L., Schreiber, E., Liebermann, R. C. & Soga, M. 1968 *Rev. Geophys.* **6**, 491–524.
Backus, G. E. & Gilbert, J. F. 1968 *Geophys. Jl R. astr. Soc.* **16**, 169–205.
Bloxham, J. & Gubbins, D. 1987 *Nature, Lond.* **325**, 511–513.
Castillo, P. R. 1988 *Eos, Wash.* **69**, 490, 491.
Creager, K. C. & Jordan, T. H. 1984 *J. geophys. Res.* **89**, 3031–3049.
Creager, K. C. & Jordan, T. H. 1986 *J. geophys. Res.* **91**, 3573–3589.
Creager, K. C. & Jordan, T. H. 1987 *Eos, Wash.* **68**, 1487.
Dahlen, F. A. 1968 *Geophys. Jl R. astr. Soc.* **16**, 329–367.
Dziewonski, A. M. 1984 *J. geophys. Res.* **89**, 5929–5952.
Dziewonski, A. M. & Anderson, D. L. 1981 *Physics Earth planet. Inter.* **25**, 297–356.
Dziewonski, A. M. & Woodhouse, J. H. 1987 *Science, Wash.* **236**, 37–48.
Dziewonski, A. M., Hager, B. H. & O'Connell, R. J. 1977 *J. geophys. Res.* **82**, 239–255.
Forte, A. M. & Peltier, W. R. 1987 *J. geophys. Res.* **92**, 3645–3679.
Forte, A. M. & Peltier, W. R. 1989 *Tectonophysics.* (In the press.)
Giardini, D., Li, X.-D. & Woodhouse, J. H. 1987 *Nature, Lond.* **325**, 405–411.
Giardini, D., Li, X.-D. & Woodhouse, J. H. 1988 *J. geophys. Res.* **93**, 13716–13742.
Grand, S. P. 1987 *J. geophys. Res.* **92**, 14065–14090.
Gubbins, D. & Bloxham, J. 1985 *Geophys. Jl. R. astr. Soc.* **80**, 695–713.
Gwinn, C. R., Herring, T. A. & Shapiro, I. I. 1986 *J. geophys. Res.* **91**, 4755–4765.
Hager, B. H. & Clayton, R. W. 1989 In *Mantle convection* (ed. W. R. Peltier). New York: Gordon & Breach. (In the press.)
Hager, B. H., Clayton, R. W., Richards, M. A., Comer, R. P. & Dziewonski, A. M. 1985 *Nature, Lond.* **313**, 541–545.
Hales, A. L. & Doyle, H. A. 1967 *Geophys. Jl R. astr. Soc.* **13**, 403–415.
Hart, S. R. 1984 *Nature, Lond.* **309**, 753–757.
Jackson, D. D. 1979 *Geophys. Jl R. astr.* **57**, 137–157.
Jeanloz, R. & Wenk, H.-R. 1988 *Geophys. Res. Lett.* **15**, 72–75.
Jordan, T. H. 1975 *Rev. Geophys. Space Phys.* **13**, 1–12.
Jordan, T. H. 1978*a* *Nature, Lond.* **274**, 544–548.
Jordan, T. H. 1978*b* *Geophys. Jl R. astr. Soc.* **52**, 441–455.
Jordan, T. H. 1988 *J. Petr.* Special lithosphere issue, pp. 11–37.
Jordan, T. H. & Lynn, W. S. 1974 *J. geophys. Res.* **79**, 2679–2685.

Lay, T. 1983 *Geophys. Jl R. astr. Soc.* **72**, 483–516.
Masters, G., Jordan, T. H., Silver, P. G. & Gilbert, F. 1982 *Nature, Lond.* **298**, 609–613.
Minster, J. B. & Jordan, T. H. 1978 *J. geophys. Res.* **83**, 5331–5354.
Mochizuki, E. 1986 *Geophys. Res. Lett.* **13**, 1478–1481.
Morelli, A. & Dziewonski, A. 1986 *Eos, Wash.* **67**, 311.
Morelli, A. & Dziewonski, A. M. 1987*a* *Nature, Lond.* **325**, 678–683.
Morelli, A. & Dziewonski, A. M. 1987*b* In *Seismic tomography* (ed. G. Nolet), pp. 251–274. D. Reidel.
Morelli, A., Dziewonski, A. M. & Woodhouse, J. H. 1986 *Geophys. Res. Lett.* **13**, 1545–1548.
Nakanishi, I. & Anderson, D. L. 1982 *Bull. Am. seism. Soc.* **72**, 1185–1194.
Nakanishi, I. & Anderson, D. L. 1983 *J. geophys. Res.* **88**, 10267–10283.
Nakanishi, I. & Anderson, D. L. 1984 *Geophys. Jl R. astr. Soc.* **78**, 573–618.
Nataf, H.-C., Nakanishi, I. & Anderson, D. L. 1984 *Geophys. Res. Lett.* **11**, 109–112.
Nataf, H.-C., Nakanishi, I. & Anderson, D. L. 1986 *J. geophys. Res.* **91**, 7261–7307.
Park, J. 1987 *Geophys. Jl R. astr. Soc.* **90**, 129–169.
Pekeris, C. L. 1935 *Mon. Not. R. astr. Soc. geophys. Suppl.* **3**, 343–367.
Richards, M. A. & Hager, B. H. 1984 *J. geophys. Res.* **89**, 5987–6002.
Ritzwoller, M., Masters, G. & Gilbert, F. 1986 *J. geophys. Res.* **91**, 10203–10228.
Ritzwoller, M., Masters, G. & Gilbert, F. 1988 *J. geophys. Res.* **93**, 6369–6396.
Romanowicz, B. 1987 *Geophys. Jl R. astr. Soc.* **90**, 75–100.
Shearer, P. M., Toy, K. & Orcutt, J. A. 1988 *Nature, Lond.* **333**, 228–232.
Silver, P. G. & Jordan, T. H. 1981 *Geophys. Jl R. astr. Soc.* **64**, 605–634.
Tanimoto, T. 1987 *Eos, Wash.* **68**, 1487, 1488.
Tarantola, A. & Valette, B. 1982*a* *J. Geophys.* **50**, 159–170.
Tarantola, A. & Valette, B. 1982*b* *Rev. Geophys. Space Phys.* **20**, 219–232.
Woodhouse, J. H. & Dahlen, F. A. 1978 *Geophys. Jl R. astr. Soc.* **53**, 335–354.
Woodhouse, J. H. & Dziewonski, A. M. 1984 *J. geophys. Res.* **89**, 5953–5986.
Woodhouse, J. H. & Dziewonski, A. M. 1986 *Eos, Wash.* **67**, 307.
Woodhouse, J. H. & Giardini, D. 1985 *Eos, Wash.* **66**, 300.
Woodhouse, J. H. & Girnius, T. P. 1982 *Geophys. Jl R. astr. Soc.* **68**, 653–673.
Woodhouse, J. H., Giardini, D. & Li, X.-D. 1986 *Geophys. Res. Lett.* **13**, 1549–1552.

Phil. Trans. R. Soc. Lond. A **328**, 309–327 (1989)
Printed in Great Britain

Long-wavelength variations in Earth's geoid: physical models and dynamical implications

By B. H. Hager and M. A. Richards†

Seismological Laboratory, California Institute of Technology, Pasadena, California 91125, *U.S.A.*

The seismic velocity anomalies resolved by seismic tomography are associated with variations in density that lead to convective flow and to dynamically maintained topography at the Earth's surface, the core–mantle boundary (CMB), and any interior chemical boundaries that might exist. The dynamic topography resulting from a given density field is very sensitive to viscosity structure and to chemical stratification. The mass anomalies resulting from dynamic topography have a major effect on the geoid, which places strong constraints on mantle structure. Almost 90% of the observed geoid can be explained by density anomalies inferred from tomography and a model of subducted slabs, along with the resulting dynamic topography predicted for an Earth model with a low-viscosity asthenosphere (*ca.* 10^{20} Pa s) overlying a moderate viscosity (*ca.* $10^{22.5}$ Pa s) lower mantle. This viscosity stratification would lead to rapid mixing in the asthenosphere, with little mixing in the lower mantle. Chemically stratified models can also explain the geoid, but they predict hundreds of kilometres of dynamic topography at the 670 km discontinuity, a prediction currently unsupported by observation. A low-viscosity or chemically distinct D″ layer tends to decouple CMB topography from convective circulation in the overlying mantle. Dynamic topography at the surface should result in long-term changes in eustatic sea level.

1. Introduction

To understand the dynamics of mantle convection and plate motions, we must probe Earth's interior by using a variety of techniques. In the past decade, seismologists have begun to produce intriguing, albeit often fuzzy, global maps of the interior of the planet. Higher-resolution studies have revealed variations in the shape and state of stress in subducted slabs. Geochemists have analysed rocks outcropping at the surface that possess isotopic patterns indicative of isolation and incubation in isolated interior reservoirs for time periods of order 1 Ga.

Long-wavelength variations in Earth's geoid provide fundamental constraints on the interior density structure and dynamics complementary to those provided by other disciplines. Geoid undulations are primarily the result of density variations associated with mantle convection. In this paper, we present physical models of the geoid that make extensive use of recent results from seismic tomography to explain nearly 90% of the variance in the observed geoid at wavelengths greater than 4000 km. A component vital to the success of these models is the inclusion of the effects on the geoid of the dynamic topography driven by mantle convection. The amplitude of this dynamic topography, and hence the total geoid, is strongly dependent upon the mechanical and compositional stratification of the mantle. Thus the geoid produced in a dynamic Earth by a given distribution of density heterogeneities, for example as inferred

† Present address: Department of Geological Sciences, University of Oregon, Eugene, Oregon 97403, U.S.A.

indirectly from seismic tomography, can be used to constrain the interior structure of our planet.

A map of the observed long-wavelength geoid (Lerch *et al.* 1983) is shown in figure 1, along with continents, plate boundaries, and active hotspots. The *ca.* 20 km hydrostatic oblateness (Nakiboglu 1982) has been subtracted to reveal the *ca.* 200 m non-hydrostatic signal of tectonic origin. The non-hydrostatic geoid is dominated by polar lows and a band of equatorial highs, the latter broken by a moderate geoid low centred south of India.

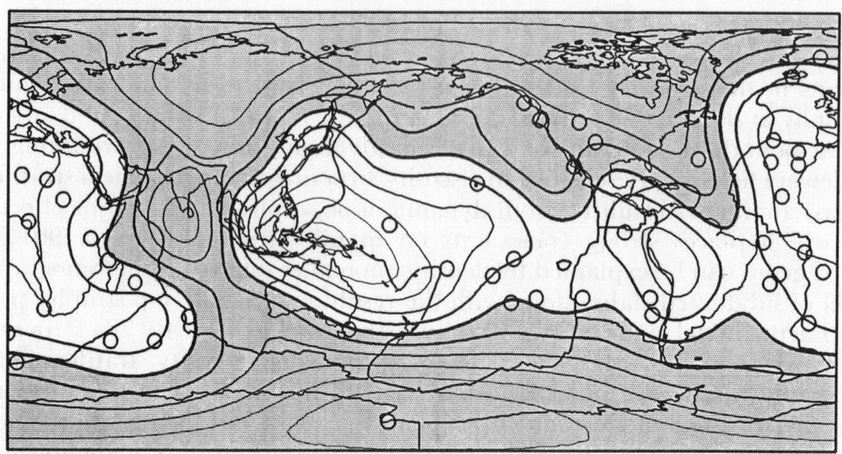

FIGURE 1. The observed long-wavelength geoid referred to the hydrostatic figure ($f = 1/299.63$). Plate boundaries and hotspots (circles) are indicated. The contour interval is 20 m and geoid lows are shaded. Cylindrical equidistant projection.

Empirical associations of the geoid with various geological and geophysical features are discussed in detail elsewhere (Richards & Hager 1988*a*). There are, in general, geoid highs associated both with convergence zones (see, for example, Kaula 1972; Chase 1979; Crough & Jurdy 1980; Hager 1984; Richards & Hager 1988*a*) and with regions having a concentration of hotspots ('hotspot provinces') (see Chase 1979; Crough & Jurdy 1980; Richards *et al.* 1988). But with the exception of convergence zones, there is little association of geoid features with plate tectonics or with continents. In particular, the spectacular association of geoid anomalies with lithospheric features so apparent at shorter wavelengths (see review by Douglas *et al.* 1987) is absent in the wavelength band shown here. Isostatic models of the geoid effects expected from density variations in the lithosphere are an order of magnitude smaller than total observed variations (Chase & McNutt 1982; Hager 1983); the primary source of the variations in figure 1 is deep.

An apparent paradox is that geoid highs are associated both with the cold, dense subducted slabs that plunge into the interior at convergence zones and with the hot, low-density mantle plumes thought to feed the hotspot volcanism at Earth's surface. This paradox has defeated successful empirical interpretation of the geoid in the context of plate tectonics and mantle convection.

Recent work in the fields of seismic tomography and fluid dynamical modelling of the mantle has made substantial progress towards resolving this apparent paradox. Studies of lower-mantle seismic velocity heterogeneity using P-wave travel-time tomography have revealed long-wavelength heterogeneities with patterns similar to the observed geoid (Dziewonski *et al.*

1977; Dziewonski 1984; Clayton & Comer 1984; Hager & Clayton 1988). In particular, regions of the lower mantle beneath the surface hotspot provinces and associated geoid highs over Africa and the central Pacific are characterized by anomalously slow mantle. If one assumes that slow mantle is hot and buoyant, and fast mantle is cold and dense, and solves Poisson's equation for the geoid predicted by the density fields inferred from lower-mantle seismic tomography, the match to the observed geoid pattern is remarkable (Dziewonski *et al.* 1977; Dziewonski 1984; Hager *et al.* 1985; Hager & Clayton 1988). Also remarkable is that the calculated geoid has the opposite sign to that observed, with geoid lows predicted over Africa and the central Pacific and geoid highs over the Poles.

This mismatch in sign is what would be expected based on the pioneering work by Pekeris (1935) over a half century ago. Pekeris pointed out that in a convecting planet, the effects of dynamically maintained topography on the geoid would be important. In fact, for a uniform-viscosity planet, the mass anomalies resulting from uplift over hot upwellings and subsidence over cold downwellings would have such a large effect on the geoid that they would reverse its sign, giving geoid highs over mantle upwellings and lows over downwellings.

Runcorn (1964, 1967) attempted to use Pekeris's theory to relate the observed geoid to plate motions and mantle convection, but had little success. The reason became clear when the theory was extended to include the effects of viscosity variation with depth, the effects of possible chemical stratification at the 670 km seismic discontinuity, and self-gravitation (Richards & Hager 1984; Ricard *et al.* 1984). Stratification of either the viscosity or composition has a major impact on the dynamic topography and resulting geoid anomalies produced for a given interior density distribution. Hence, if the density distribution can be estimated by geophysical observations or models, the geoid provides a sensitive probe of mantle structure.

Hager *et al.* (1985) showed that most of the geoid at wavelengths longer than 10000 km can be explained as the result of heterogeneities in the lower mantle inferred from seismic tomography, along with the associated dynamic topography. Using simple two-layer models to parametrize the viscosity distribution, they obtained a good agreement with the observed geoid for a model in which the lower mantle had a viscosity a factor of 10 greater than that of the upper mantle.

Hager (1984) showed that the geoid in the wavelength 4000–10000 km is dominated by the signature of subducted slabs. He also used a simple two-layer parametrization of viscosity to calculate model geoids; these models required a lower-mantle viscosity a factor greater than 30 larger than that of the upper mantle to match the observed geoid.

In this paper, we extend these models to include a more detailed and realistic parametrization of the viscosity structure of the mantle. This parametrization refines the structure of the upper-mantle to include a high-viscosity lid, a low-viscosity asthenosphere, and an increase in viscosity through the transition zone. There is increasing evidence that the D″ layer at the base of the mantle is anomalous in viscosity and/or composition, and we include these possible effects. We also include the effects of heterogeneity inferred from surface-wave studies of upper-mantle structure. We are now able to explain about 90% of the variance in the observed geoid in the wavelength band represented by spherical harmonic degrees 2–9.

In the following section, we briefly review the fluid mechanics of geoid anomalies in a convecting planet. We derive a Green function approach whereby the gravitational effects for any interior density field, including the effects of dynamic topography, can be calculated for a given model of viscosity and compositional layering of the mantle. We then present model geoids calculated for a few particular Earth models for density fields inferred from lower-mantle P-wave tomography, from upper-mantle surface-wave tomography, and from a density model for subducted slabs.

Successful models have a substantial viscosity increase (*ca.* 300) from the asthenosphere to the lower mantle. The geoid is sensitive to relative variations in viscosity, but it is not sensitive to the absolute level of the viscosity. The heat advected by the flow associated with the inferred density contrasts is, however, sensitive to the absolute viscosity, from which a lower bound to lower-mantle viscosity of *ca.* 3×10^{22} Pa s, an order of magnitude higher than that inferred by some recent studies of postglacial rebound, can be inferred (O'Connell & Hager 1984; Hager & Clayton 1988).

We discuss the geodynamical consequences of this viscosity stratification. These include a stratification in the style of mantle convection, with the upper mantle much more quickly mixed than the lower mantle. The slow circulation time for the lower mantle, compared with the faster timescale for the reorientation of plate boundaries, could explain why there is not a more obvious association of the geoid with surface tectonics, as well as the long residence of isotopic reservoirs. The viscosity stratification could also explain the shape and state of stress and seismicity distribution observed for subducted slabs (Vassiliou *et al.* 1984; Gurnis & Hager 1988). Consideration of recent constraints on the topography of the core–mantle boundary (CMB) suggests that there may be a layer of molten silicate at the top of the core.

2. Fluid mechanics of geoid anomalies

(*a*) *Physical effects*

The geoid is the equipotential surface corresponding to mean sea level. Variations in δN, the height of the geoid, relative to that expected for a hydrostatic planet, result from variations in gravitational potential δV by

$$\delta N = \delta V(a)/g(a), \tag{1}$$

where $g(a)$ is the gravitational acceleration at the Earth's mean radius a. Variations in gravitational potential are caused by non-hydrostatic variations in density $\delta\rho$. The two are related by Poisson's equation

$$\nabla^2 \delta V = -4\pi\gamma\,\delta\rho. \tag{2}$$

Here γ is the gravitational constant.

To determine the total gravitational potential, Poisson's equation must be integrated over the volume of the Earth. Convection in the mantle results in dynamic topography (Pekeris 1935). (Mid-oceanic rises, which result from cooling of the upper thermal boundary layer, and deep sea trenches are familiar examples of convectively maintained topography.) It is crucial to include the density anomalies resulting from this dynamic topography in determining the total geoid anomaly resulting from a given internal density field. At long wavelengths, the total mass displaced by dynamic topography at the top and bottom of a convecting system is comparable to the mass excess of the internal anomalies (Richards & Hager 1984). A sort of 'dynamic isostasy' approximately holds. The gravitational effects of the interior density field and the resulting dynamic topography are comparable; the total geoid is a small difference between relatively large numbers.

It is, unfortunately, difficult to observe directly the amplitude and pattern of dynamically maintained topography resulting from deep-seated density variations. Topography generated by density variations within the lithosphere, including crustal-thickness variations, dominates Earth's surface relief. There are indications that the dynamic surface topography is of order 1 km. For example, over geologic time, continental shields might be expected to be reduced by erosion to similar hypsometries. The African Shield stands hundreds of metres higher than other shields, presumably as the result of recent dynamic uplift. The ridge south of Australia is over 1000 m deeper than the average ridge crest, while the central Pacific is elevated up to 1 km (Crough 1983; Cazenave *et al.* 1988). However, although qualitative analysis indicates the existence of dynamic topography, the density structure of the lithosphere is not yet known well enough to remove its effects quantitatively. Instead we use fluid-dynamical models of mantle flow to calculate the dynamic topography.

Recently, geodetic, magnetic, and seismological observations have been used to infer the topography of the CMB. Because the CMB is hot, it is likely that any topography there is dynamically maintained. Gwinn *et al.* (1986) inferred from nutation amplitudes given by VLBI (very-long-baseline interferometry) measurements that the CMB has an excess ellipticity of *ca.* 500 m. Hide (1986) estimates that topography on the CMB is of order of a few hundred metres from models of the coupling of flow in the core to changes in length of day. Models of CMB topography from seismic tomography range from 2–12 km (Morelli & Dziewonski 1987; Gudmundsson *et al.* 1987), however. These discrepancies are discussed below.

(*b*) *Governing equations, assumptions and method of solution*

Over geologic timescales, the rocks in the Earth's mantle respond to stresses by slow, creeping flow. Inertial forces are negligible and the equilibrium equation becomes

$$\nabla \cdot \boldsymbol{\tau} + \rho \boldsymbol{g} = 0, \tag{3}$$

where $\boldsymbol{\tau}$ is the stress tensor, ρ the density, and $\boldsymbol{g}$ the gravitational acceleration.

The stress tensor $\boldsymbol{\tau}$ is related to $\boldsymbol{\varepsilon}$, the strain rate tensor, by the constitutive law:

$$\boldsymbol{\tau} = -p\boldsymbol{I} + 2\eta\boldsymbol{\varepsilon}, \tag{4}$$

where $\boldsymbol{I}$ is the identity matrix, p is the pressure, and η is an effective viscosity, in general a function of temperature, pressure, composition, strain rate, and total strain. For the models presented here, for the sake of mathematical tractability, we assume η to be spherically symmetric. This assumption is critically examined in Richards & Hager (1988*b*). They find that for half wavelengths on the order of the thickness of the mantle or greater, the effects of radial variations of viscosity are more important than the effects of lateral variations.

Finally, as discussed in Richards & Hager (1984), we make the approximation that the flow is incompressible. Zhang & Yuen (1987) have addressed this issue quantitatively. Their results show that although compressibility has some effect, this effect is small compared to effects of, for example, radial variations in viscosity. For an incompressible fluid, the continuity equation becomes

$$\nabla \cdot \boldsymbol{v} = 0 \tag{5}$$

with $\boldsymbol{v}$ the flow velocity.

Under the assumption that the viscosity is spherically symmetric, the coupled set of equations (2)–(5) can be solved analytically. A full discussion is given in Hager & Clayton (1988), where an updated (O'Connell *et al.* 1984), somewhat cleaner derivation of the equations used by

Richards & Hager (1984) is presented. The approach is to separate the radial and azimuthal variations of all variables by expanding azimuthal variations in terms of scalar or vector spherical harmonics and to utilize the orthonormality properties of the spherical harmonic basis functions to isolate the radial dependence for each spherical harmonic degree and order. The radial dependence is then solved using the propagator matrix technique (Gantmacher 1960).

The flow and resulting stresses and surface deformations are driven by interior density contrasts specified *a priori*, for example, from a density distribution inferred from seismic tomography. For mathematical simplicity, volumetric density contrasts are collapsed to form surface-mass anomalies spaced at 100 km intervals or less.

The physical boundary conditions applied are continuity of velocity and tractions across the CMB, the surface, and any interior compositional boundaries that are included in a given model (e.g. 670 km discontinuity, top of D″). Because the propagator matrix technique assumes spherical boundaries, these physical boundary conditions are analytically continued from the physical (deformed) boundaries to the mathematical (spherical) boundaries by using a first-order Taylor's series expansion. The result is that the mathematical boundary conditions become continuity of velocity and shear traction, with a jump condition in radial normal stress τ_{rr}:

$$\tau_{rr}]_{-}^{+} = \Delta\rho g\,\delta h. \tag{6}$$

Here δh is the dynamic topography of the boundary and $\Delta\rho$ is the density contrast across the boundary. At the surface and CMB, the radial velocity and shear traction vanish. At an intermediate boundary in composition, the radial velocity vanishes, while the shear tractions are non-zero but continuous.

For a given τ_{rr}, δh is inversely proportional to $\Delta\rho$. We will see later that at the surface and CMB, where $\Delta\rho$ is large, the inferred δh is small compared to the depth of the convecting mantle. For models with chemical stratification at 670, δh is large and the approximation discussed above is not very good.

(*c*) *Kernels*

In this paper, we focus on the values of the gravitational potential at the surface, along with the dynamic topography at the surface, the CMB, and any interior compositional boundaries that might exist. Values are predicted for various models of the interior density field. It is instructive to construct response functions for these quantities for a suite of models of Earth structure. These response functions, or kernels, are expressed for an interior density contrast of unit amplitude of a given spherical harmonic degree l and order m at a specified radius r, i.e. they are mixed spectral–spatial Green's functions. The calculation is linear, given the assumptions stated above, so the total response is obtained by convolving these kernels with a particular distribution of density contrasts.

Because we assume azimuthal symmetry of viscosity, the kernels depend upon spherical harmonic degree l, but not upon order m (Hager & Clayton 1988). We define normalized potential kernels $G^{l}(r)$, normalized surface displacement kernels $A^{l}(r)$, normalized CMB displacement kernels $C^{l}(r)$, and normalized chemical discontinuity topography kernels $D^{l}(r)$ such that for a spherical harmonic coefficient and density contrast $\delta\rho^{lm}(r)$

$$\delta V_{(a)}^{lm} = \frac{4\pi\gamma a}{2l+1}\int_{c}^{a} G^{l}(r)\,\delta\rho^{lm}(r)\,\mathrm{d}r, \tag{7}$$

$$\delta a^{lm} = \frac{1}{\Delta\rho_a} \int_c^a A^l(r)\, \delta\rho^{lm}(r)\, \mathrm{d}r, \tag{8}$$

$$\delta c^{lm} = \frac{1}{\Delta\rho_{\mathrm{CMB}}} \int_c^a C^l(r)\, \delta\rho^{lm}(r)\, \mathrm{d}r, \tag{9}$$

$$\delta d^{lm} = \frac{1}{\Delta\rho_d} \int_c^a D^l(r)\, \delta\rho^{lm}(r)\, \mathrm{d}r. \tag{10}$$

Here δa^{lm} is the dynamic surface topography (at $r = a$), δc^{lm} the dynamic topography of the CMB (at $r = c$), and δd^{lm} the dynamic topography of the interior chemical discontinuity (at $r = d$). With the normalization for the potential kernel, if the Earth were static, with no dynamic topography, $G^l(r) = (r/a)^{(l+2)}$.

Geoid kernels for two-layer models have been discussed in Hager (1984) and Hager *et al.* (1985). Topography kernels for two-layer models were presented in Richards & Hager (1984). An extensive discussion of the effects of adding more layers of varying viscosity is given in Hager & Clayton (1988). For reasons of space limitations, we present here kernels for only a limited set of models that are successful in matching the observed geoid.

Recognizing that there are many ways in which a model could be parametrized, we feel it important to explain our particular choice. First, the propagator matrix formulation forces us to parametrize the viscosity as a series of layers, each with constant viscosity. We have attempted to keep the number of layers small, yet at the same time we have been influenced by *a priori* expectations as to where the viscosity might be expected to change substantially.

Probably the major rheological stratification in the mantle is the contrast between the strong lithosphere and the underlying asthenosphere. While in plate interiors, the lithosphere has effectively an infinite viscosity, viewed on a global scale, the lithosphere does deform, resulting in relative motions between the plates. While the effective viscosity of the lithosphere is certainly heterogeneous, with deformation concentrated at presumably weak plate boundaries, our mathematical technique forces us to ignore this heterogeneity. We model the lithosphere as a viscous fluid of thickness 100 km; operationally, this viscosity can be thought of as the average stress divided by the average strain rate (Kaula 1980; Hager & O'Connell 1981). There is, of course, a tradeoff between the thickness of the lithosphere and its effective viscosity.

Increases in pressure with depth lead to continuous increases in viscosity, although these are partially offset by increases in temperature along the adiabat (Ashby & Verrall 1977) or reversed in thermal boundary layers. Localized phase changes and compositional changes are expected to lead to discontinuous changes in viscosity that can be large (Sammis *et al.* 1977). For this reason, we place boundaries between our viscous layers at the seismic discontinuities at 400 km and 670 km depth and at the top of D″. There will be a tradeoff between the viscosity jump at layer boundaries and continuous increases in viscosity within layers (Revenaugh & Parsons 1987). We search a model space in which the relative viscosity in a given layer can vary only in steps of $10^{0.5}$.

In our previously published models of the geoid, we have not included the effects of structure at the base of the lower mantle, primarily because it is possible to obtain excellent fits to the geoid without the added model complexity that results from including an additional layer. We include it here for several reasons. First, D″ forms the boundary region between the lower mantle and the molten core; globally averaged seismic models (Dziewonski & Anderson 1981) show a region of decreased velocity gradient in D″. D″ is almost certainly a region of increased temperature gradient, i.e. a hot thermal boundary layer, and hence is expected to have a low viscosity. More-detailed regional models (Lay & Helmberger 1983*a*, *b*; Young & Lay 1987) show a sharp boundary, so D″ may well be distinct in composition as well. Second, there is now great interest in constraining the properties of the CMB. The predicted dynamic topography at the CMB is a strong function of the structure of D″.

Kernels for several models that allow us to explain most of the variance in the long-wavelength geoid are shown in figures 2–4. For these models, the first letter refers to the style of convection: 'W' for whole mantle and 'C' for a mantle chemically stratified at 670 km depth. The second letter refers to D″: 'L' for low viscosity and 'C' for chemically distinct (also with low viscosity). Only relative viscosity variations affect the geoid. The distribution of relative viscosity η with depth is shown in the figures. For example, for our preferred model WL, the distribution is as follows: 0–100 km depth, $\eta = 1$; 100–400 km, $\eta = \frac{1}{30}$; 400–670 km, $\eta = 1$; 670–2600 km depth, $\eta = 10$; 2600 km depth – CMB, $\eta = \frac{1}{10}$.

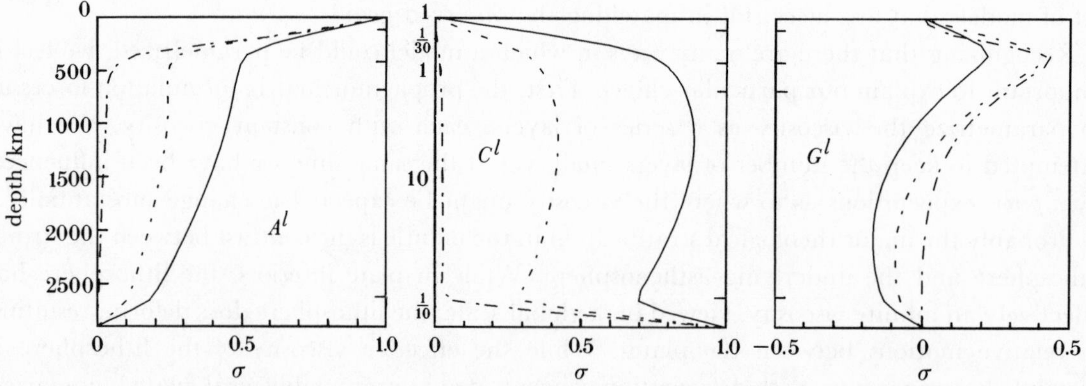

FIGURE 2. Normalized kernels for model WL, showing the effects of a surface mass σ placed at a given depth in the mantle for $l = 2$, 4, and 8. The masses displaced by surface deformation and by deformation of the CMB are normalized by dividing by $(-\sigma)$. A value of unity implies perfect dynamic compensation. The geoid kernel, which includes the effects of the dynamic topography as well as σ, is normalized by the potential due to a mass σ at the surface of the Earth. The solid line is for degree two, short dashes for degree four, and long dashes for degree eight.

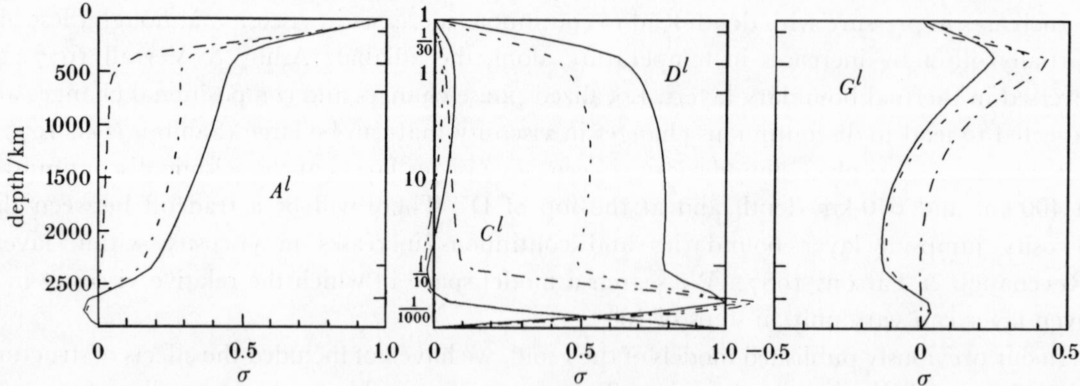

FIGURE 3. As in figure 2, but for model WC, which differs from model WL by having a chemically distinct D″ layer with a low-viscosity layer above it in the depth range 2300–2600 km to simulate the effects of an additional hot thermal boundary layer.

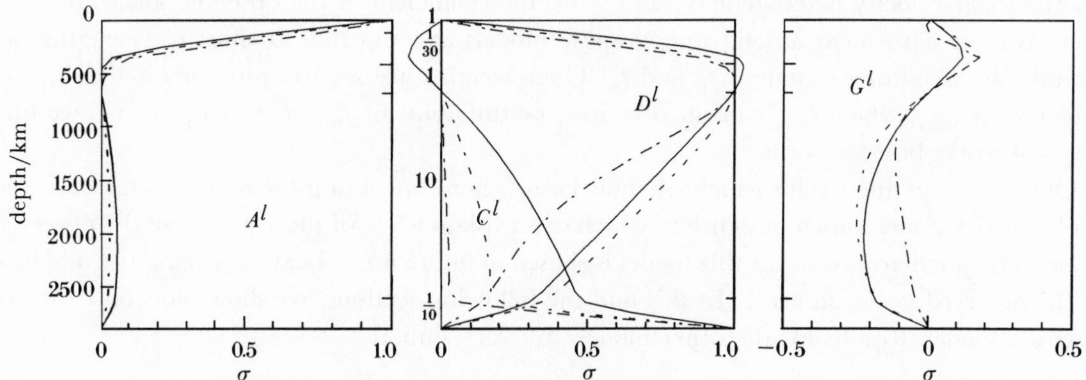

FIGURE 4. As in figure 2, but for model CL, which differs from model WL by having a boundary in chemical composition at the 670 km discontinuity.

3. Dynamical models of the geoid

(a) *Forward modelling with inversion for parameters*

Our previous modelling has shown that density contrasts associated with lower-mantle heterogeneity inferred from seismic tomography under the assumption that these velocity variations are thermal in origin (Hager *et al.* 1985) and upper-mantle density contrasts associated with subducted slabs (Hager 1984) can explain much of the long-wavelength undulation in the observed geoid. There are, however, substantial uncertainties associated with these models. For example, the ratio of density perturbation to velocity perturbation for the lower mantle is not known *a priori*. The actual mass anomalies associated with subducted slabs are relatively poorly constrained. There is a possible tradeoff between these parameters and the viscosity structure that best explains the observed geoid. Large-amplitude variations in upper-mantle shear-wave velocities have recently been mapped (Masters *et al.* 1982; Woodhouse & Dziewonski 1984; Nataf *et al.* 1984, 1986; Tanimoto & Anderson 1984; Tanimoto 1986; Grand 1987). These velocity anomalies could have substantial density contrasts associated with them and therefore have an effect on the geoid and dynamic surface topography. Although the various tomographic inversions tend to agree at long wavelengths, this agreement disappears at shorter wavelengths, opening the question of reliability of the inferences of mantle structure.

To investigate some of the tradeoffs and uncertainties associated with modelling the geoid, we have chosen to fit the geoid by solving the following set of linear equations (Hager & Clayton 1988):

$$R s^{lm} + P f_{\mathrm{p}} V_{\mathrm{p}}^{lm} + S f_{\mathrm{s}} V_{\mathrm{s}}^{lm} = g^{lm}. \tag{11}$$

The equations are solved for R, the density contrast of mature subducted slabs (assumed distributed over a thickness of 125 km and dipping at 60°), P, the ratio of density perturbation to velocity perturbation for lower-mantle compressional-wave velocity anomalies, and S a similar ratio for upper-mantle shear-wave velocity anomalies, all assumed to be constants independent of radius. In this equation, s^{lm} is the geopotential coefficient obtained from (7) using the subducted slab model described below and V_{p}^{lm} is a similar integral involving the convolution of the potential kernel with the spherical harmonic expansion of P-wave velocity heterogeneity in the lower mantle. V_{s}^{lm} is obtained in the same way using a model of upper-mantle shear-velocity heterogeneity, and g^{lm} is the coefficient of the observed geoid. Because there is more agreement among tomographic models at long than at short wavelengths, we include the weighting functions f_p and f_s. These weights are set to unity for $l \leqslant l_{max}$ and to zero for $l > l_{max}$, where l_{max} can, in principle, be different for f_p and f_s. In practice, we find $l_{max} = 4$ works best for both.

Of the viscosity models for which we show kernels here, we obtain the best fit to the observed geoid for the model shown in figure 2, which can explain 87% of the variance of the observed geoid. The predicted geoid for this model is shown in figure 5*a*; it bears a striking resemblance to the observed geoid, figure 1. In this and the following sections, we show plots only for this viscosity model. Results for the other models are very similar.

(*b*) *Individual contributions*

The model geoids shown (discussed separately in the next sections) were computed simultaneously by solving (11). The geoid contributions from the three sources are not totally orthogonal and there is some tradeoff among the parameters in (11), as discussed below in (*c*). However, this tradeoff is not so large that the individual contributions are qualitatively affected by it. The individual contributions do not change very much if, for example, they are solved for sequentially, as in our previously published results (Hager 1984; Hager *et al.* 1985).

(i) *Lower-mantle contributions to the geoid*

The contribution to the geoid calculated with the lower mantle P-wave tomographic inversion of Clayton & Comer (1984) is shown in figure 5*b*. As discussed in more detail in Hager & Clayton (1988), although the tomographic inversion is a global one, the ray coverage in the upper mantle and near the base of the mantle for this and other P-wave inversions is very poor. For this reason, we evaluate the integral in equation (7) only between the depths of 800–2800 km. We also set l_{max} at 4; above this value, the overall fit to the geoid degrades slightly. The density perturbation to velocity perturbation P has a value of 0.3 $(\mathrm{Mg\ m^{-3}})/(\mathrm{km\ s^{-1}})$.

The agreement between this predicted long-wavelength geoid and the observed geoid is striking. Both are dominated by polar lows and equatorial highs. Both encompass most of the hotspots within their highs. In fact, the correlation between the degree two component of the spherical harmonic distribution of hotspots and this predicted geoid is almost perfect ($r = 0.99$), suggesting that surface hotspots and lower-mantle thermal structure are strongly coupled (Hager *et al.* 1985; Richards & Hager 1988*a*; Richards *et al.* 1988).

The longest-wavelength features, which contribute most of the power to the geoid, appear to be explained primarily by structure in the lower mantle. The predicted geoid does not have the highs over the Andes and the western Pacific seen in the observed geoid. These can be explained by density contrasts associated with subducted slabs in the upper mantle.

(ii) *Subducted-slab contributions to the geoid*

The foci of deep earthquakes at subduction zones mark positions of the cold, central cores of subducted slabs. If we knew the density contrasts associated with these subducted slabs, we could calculate the associated geoid anomalies for a given model of Earth structure using (7). Unfortunately, whereas seismicity constrains the location of subducted slabs to some extent, the associated density contrasts are not well constrained. Although the integrated density contrast between subducted slabs and the asthenosphere in the shallow mantle is constrained by the subsidence of the lithosphere with age as it moves away from mid-oceanic ridges, it is not well constrained at depth. Complicating factors include the effects of phase changes (see, for example, Schubert *et al.* 1975; Anderson 1987), the effects of slab extension and compression (Isacks & Molnar 1971), and the likelihood that slab material is present even when there is no seismicity (see Hager (1984) and Hager & Clayton (1988) for an extended discussion). Thus, in calculating the geoid anomaly associated with subducted slabs, the distribution of mass anomalies, as well as the mantle structure, must be assumed. The validity of assumed mass distributions can then be tested by comparing the model geoid predicted to the observed.

For simplicity, we assume that the surface mass anomaly associated with subducted lithosphere is conserved with depth. We ignore slab extension and compression and assign all slabs a dip of 60°. Although seismicity cuts off at some depth for each subduction zone, probably as the result of warming of the core of the slab beyond the temperature at which brittle failure can occur for the relevant phase, the mass anomaly associated with the slab probably extends beneath the deepest earthquakes. We assume that all slabs penetrate to the same depth. For models with flow permitted into the lower mantle, we choose the lower limit in (7) to be at a depth of 800 km to avoid double counting with the lower-mantle tomographic model. For chemically stratified models, we choose 670 km as the natural lower limit for slab penetration, in accord with the small boundary deformation approximation in §2*b*. For slabs where seismicity is confined to depths shallower than 300 km, probably as the result of shallower reheating of young or slowly subducting lithosphere, we assume that the mass anomaly and depth of seismicity are related and use a mass anomaly a factor of two smaller. In the future, more detailed tomographic inversions for slab structure (Zhou & Clayton 1987) should allow us to remove many of these assumptions.

The geoid predicted for the Earth model shown in figure 2 is shown in figure 5*c*. The slab model reproduces the geoid highs associated with subduction zones. The predicted geoid is quite similar to the model obtained with a simpler two-layer viscosity model by Hager (1984), but in this case, the same viscosity model that fits the lower-mantle density model is able to fit the geoid resulting from subducted slabs. The factor R in (11) has a value of 0.10 Mg m^{-3}.

(iii) *Upper-mantle contributions to the geoid*

Inversions for upper-mantle shear-wave velocity heterogeneities reveal large variations that, if associated with large density variations, might have an observable effect on the geoid. There is a fundamental problem in comparing the results of models from surface-wave tomography and the geoid, however. Geoid kernels are zero at the surface and tend to peak at depth. On the other hand, surface waves are most sensitive to near-surface structure, and models tend to smear the effects of near-surface variations over extensive depth ranges (Tanimoto 1986). (This smearing effect is probably why subducted slabs are not visible in the uppermost mantle using surface-wave tomography; they are camouflaged by the overlying, slow, arc region. It is because of this camouflaging effect that it is necessary to include subducted slabs separately as an additional source of density contrast.) Thus the noisiest (deep) regions of the surface-wave tomographic models are highlighted by the geoid kernels.

Probably for this reason, although all the recent upper-mantle seismic models have striking similarities, we have had the most success matching the observed geoid using the longest

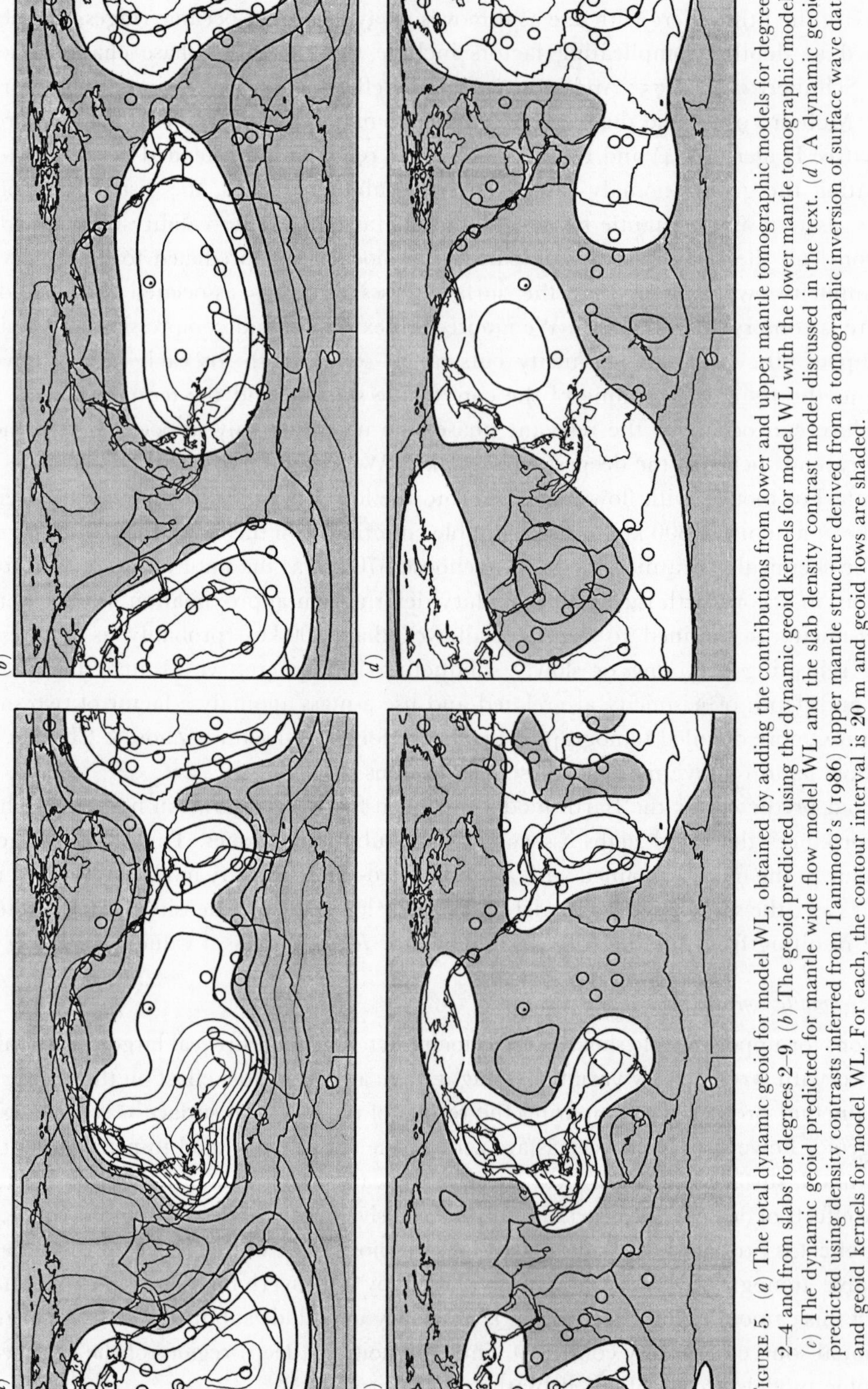

FIGURE 5. (*a*) The total dynamic geoid for model WL, obtained by adding the contributions from lower and upper mantle tomographic models for degrees 2–4 and from slabs for degrees 2–9. (*b*) The geoid predicted using the dynamic geoid kernels for model WL with the lower mantle tomographic model. (*c*) The dynamic geoid predicted for mantle wide flow model WL and the slab density contrast model discussed in the text. (*d*) A dynamic geoid predicted using density contrasts inferred from Tanimoto's (1986) upper mantle structure derived from a tomographic inversion of surface wave data and geoid kernels for model WL. For each, the contour interval is 20 m and geoid lows are shaded.

wavelengths of Tanimoto's (1986) upper-mantle model, a model which does not permit the large rapid radial variations of some other models. The predicted model, with l_{max} set to 4, is shown in figure 5*d*. The main features in this model geoid are the degree-two high over the western Pacific and the Atlantic, associated with fast material in the transition zone (Masters *et al.* 1982), and lows over North America, Antarctica, and the Carlsberg Ridge. The parameter S in (11) has a value of 0.05 (Mg m^{-3})/(km s^{-1}).

(iv) *The residual geoid and postglacial rebound contributions to the geoid*

The residual geoid, obtained by subtracting the model geoid in figure 5*a* from the observed geoid, is shown in figure 6. It is characterized by a number of geophysically interesting features. These include residual geoid highs over most of the more prominent hotspots, a phenomenon addressed in Richards *et al.* (1988). Hudson Bay and West Antarctica, loci of recent major deglaciation, remain as residual lows; indeed, the signature of delayed rebound has been recognized and discussed before (Peltier & Wu 1982). Including the gravitational effects of incomplete postglacial rebound allows over 91 % of the geoid variance to be explained (Hager & Clayton 1988).

FIGURE 6. The residual geoid obtained by subtracting the model geoid in figure 5*a* from the observed geoid in figure 1. The contour interval is 20 m and geoid lows are shaded.

(*c*) *Discussion of models and tradeoffs*

We have shown results for one particular Earth model. However, as can be seen from figures 2–4, the kernels for a number of significantly different structural models are similar. For example, the positive geoid kernels required by subducted slabs for the upper mantle can occur for either mantle-wide flow models (e.g. WL and WC) or models where the 670 km discontinuity is a chemical boundary (CL). The main difference is in the amplitude of the kernels. Also, similar geoid kernels can be obtained for D″ parametrized as a low-viscosity thermal boundary layer (WL) or a layer of distinct composition (WC); quite good fits to the geoid can also be obtained by ignoring D″ altogether (Hager & Clayton 1988).

Results for a number of models are given in table 1, where values of R, P and S, along with uncertainties and covariances, are given. Also given are the percentages of the variance in the observed geoid explained by the model. Results are given for the three models with kernels shown in figures 2–4, for a hybrid model WL′ that uses slightly different near-surface viscosity models for the slabs and upper-mantle contributions, and for three models, discussed in Hager & Clayton (1988), that do not include D″ (models W, W′, and C).

Model WL′ uses a 'lithospheric' viscosity that is one tenth that in WL for calculating the geoid due to slabs and $10^{-0.5}$ that of WL for the upper-mantle surface-wave model, for reasons discussed below and in Hager & Clayton (1988). Models W, W′, and C are like WL, WL′, and CL, but lack the D″ layer.

The lithosphere at subduction zones is cut completely through by faults, with the strain associated with plate convergence highly localized. One of the major sources of buoyancy driving plate motions is, of course, localized there as well in the form of high-density subducted slabs. We view models WL′ and W′, which have weaker 'lithospheric' layers used in the geoid models for subducted slabs, as regionalized Earth models more appropriate for subduction zones (Hager & Clayton 1988). In these models, we calculate the geoid signatures of the lower mantle, the upper mantle, and subducted slabs, all with different (spherically symmetric) models for the uppermost layer, or lid. Finite-element calculations that include a lower viscosity for the lid near subduction zones perturb the geoid kernels in a similar way to a global low-viscosity lid (Richards & Hager 1988*b*).

All of the models in table 1 give an adequate fit to the geoid. Given the assumptions involved, there is probably not much significance to the small differences among them in variance reduction. Other geophysical considerations must be used to discriminate among them.

TABLE 1. MODEL RESULTS

model	WL	WL′	WC	CL
$R/(\mathrm{Mg\ m^{-3}})$	0.10 ±0.01	0.071±0.007	0.095±0.011	0.33±0.03
$P/(\mathrm{Mg\ m^{-3}})/(\mathrm{km\ s^{-1}})$	0.29 ±0.02	0.19 ±0.01	0.40 ±0.03	0.17±0.01
$S/(\mathrm{Mg\ m^{-3}})/(\mathrm{km\ s^{-1}})$	0.046±0.006	0.047±0.006	0.046±0.007	0.10±0.01
$10^4\ C_{RP}$	−0.15	−0.10	−0.42	−0.49
$10^5\ C_{RS}$	−0.17	−0.31	−1.1	−1.7
$10^4\ C_{PS}$	−0.15	−0.10	−0.30	−0.55
CMB non-hydrostatic ellipticity/m	880	440	80	230
% variance explained	87	87	81	86

model	W	W′	C
$R/(\mathrm{Mg\ m^{-3}})$	0.16±0.02	0.064±0.006	0.36±0.09
$P/(\mathrm{Mg\ m^{-3}})/(\mathrm{km\ s^{-1}})$	0.49±0.03	0.48 ±0.02	0.35±0.03
$S/(\mathrm{Mg\ m^{-3}})/(\mathrm{km\ s^{-1}})$	0.11±0.02	0.041±0.005	0.15±0.02
$10^4\ C_{RP}$	−0.48	−0.94	−1.4
$10^5\ C_{RS}$	−0.13	−0.50	−7.1
$10^4\ C_{PS}$	−0.65	−0.20	−0.60
CMB non-hydrostatic ellipticity/m	2000	2000	1800
% variance explained	84	89	82

There are substantial differences among the model predictions in other respects, however. The models that ignore D″ all have too large an excess ellipticity associated with the CMB to be in accord with the value of *ca.* 500 m inferred from nutation observations (Gwinn *et al.* 1986). Models WL, WL′, and CL all have excess ellipticities on the order of 500 m, whereas WC has a much smaller excess ellipticity. Because most of D″ is not sampled in the tomography model used, and the CMB topography kernels are large in D″, an exact match should not be expected.

There is also substantial variation in the constants *R*, *P* and *S* for the different models, variation that is much greater than the formal statistical estimates for any given model. This is an illustration of a common phenomenon in geophysical modelling; the most important uncertainties are the result of model parametrization, and formal error estimates are not very meaningful.

The most stable parameter is *P*, the ratio of density to P-wave velocity variation in the lower mantle, which varies from 0.2 to 0.5 (Mg m^{-3})/(km s^{-1}). These values are reasonable based upon limited laboratory measurements of upper-mantle minerals; few direct measurements of lower mantle minerals are available. The value of this ratio obtained from differencing values at the top and bottom of the lower mantle in Earth model PREM (Preliminary Reference Earth Model) (Dziewonski & Anderson 1981) is 0.4, within the range calculated.

For subducted slabs, *R* ranges by more than a factor of five, from 0.064, the value obtained from the subsidence of oceanic lithosphere with age (Hager & Clayton 1988), to 0.36 Mg m^{-3}. The high value could be indicative of changes in phase (Anderson 1987). Compressive shortening of the slab as it approaches the 670 km discontinuity would also tend to increase the inferred value of *R*.

There is also substantial variation in *S*. The lower values are consistent with variations in shear-wave velocity in the upper mantle being in part of compositional origin or caused by relaxation effects. The larger values, when combined with the observed amplitude of velocity variations, would require temperature variations of *ca.* 1000 K (see Hager & Clayton 1988 for discussion).

The relatively small covariance between slab density *R* and upper-mantle factor *S* provides justification for considering slabs and upper-mantle tomography separately. We feel that *a priori* modelling is more reliable in estimating slab structure than surface-wave tomography.

4. Geophysical implications

(*a*) *Viscous or chemical stratification?*

Although our preferred model allows flow through the 670 km seismic discontinuity, we cannot discard the hypothesis of a compositional barrier at that depth based on geoid models, as acceptable fits can be obtained for this type of model. The situation is similar to the state of stress in deep subducted slabs, where both models that have a compositional barrier at 670 km depth and models that have uniform chemistry but a substantial increase in viscosity at this depth can explain the observations (Vassiliou *et al.* 1984).

There are other observations that favour flow through the 670 km discontinuity. The association of low-velocity lower mantle with regions of extensive hotspot activity at the surface is suggestive of mantle-wide flow. Travel-time anomalies from deep earthquakes interpreted in terms of deeply extending high-velocity anomalies are also suggestive of transport of at least some material across the 670 km discontinuity (Jordan 1977; Creager & Jordan 1984, 1986). However, both these examples of apparent continuity of temperature fields across the upper-mantle–lower-mantle boundary could conceivably be the result of thermally coupled convection in a layered system.

The best way of discriminating between these two types of convection might be by using the dynamic topography of the 670 km discontinuity predicted to exist if the mantle were chemically stratified. Given the large density contrasts inferred for subducted slabs in this type of model to explain the observed geoid, the dynamically maintained topography would be several hundred kilometres or more in the vicinity of subducted slabs (Hager & Raefsky 1981; Christensen & Yuen 1984). The deepest earthquakes are not resolvably deeper than the average depth of the 670 km discontinuity, suggesting that a nearly isobaric phase change, rather than a chemical barrier to flow, shuts off seismicity there. Detailed observations of phases converted at the '670' beneath the Tonga slab show no indication of substantial topography (Richards & Wicks 1987). The great success that we have had in matching the observed geoid by including the effects of dynamic topography gives us confidence in the general principles involved. The lack of topography observed on the 670 km discontinuity seems difficult to reconcile with its being a compositional barrier to convection. One possible way out would be if an approximately isobaric phase change at *ca.* 670 km depth camouflages the dynamic topography there.

Although the geoid places constraints upon relative variations in viscosity, it is insensitive to the absolute value of mantle viscosity, so long as the viscosity is high enough that the Reynolds number of the flow is negligible. But the velocity of the flow associated with a given density field is, of course, inversely proportional to the absolute viscosity. O'Connell & Hager (1984) (see also Hager & Clayton 1988) used the heat flux inferred to result from flow driven by the density fields used in geoid models to bound the viscosity of the lower mantle. So that the flow does not advect more heat than is observed to pass through Earth's surface, they argued that the lower mantle viscosity must be 3×10^{22} Pa s or more. This value of mantle viscosity is consistent with the gravitational signature of postglacial rebound discussed in §3*b*iv.

Several other lines of evidence support the idea of a substantial viscosity increase in the lower mantle. Gurnis & Hager (1988) have recently investigated fluid-dynamical models of slabs sinking through a mantle with a viscosity structure similar to that inferred from geoid modelling. They found that in addition to matching the observed state of stress, slabs in their models often showed kinks as they passed into the lower mantle similar to those inferred from analysis of seismic travel times (Jordan 1977; Creager & Jordan 1984, 1986; Zhou & Clayton 1987).

A substantial increase in the viscosity of the lower mantle results in a substantial difference between the upper and lower mantle in the style of convection. So that stresses are continuous across the boundary, strain rates must vary inversely with the viscosity. Thus the upper mantle should be much more thoroughly sheared and mixed than the lower mantle, in accord with geochemical inferences.

There is also geophysical evidence supporting the concept of slow flow velocities, presumably the result of high viscosity, in the lower mantle. Grand (1987) interpreted a high-velocity anomaly beneath North America in the lower mantle in his regional tomographic inversion as a fossil remnant of the Farallon plate, suggesting a residence time in the lower mantle of over 100 Ma. This inferred slow flow is consistent with Anderson's (1982) and Chase & Sprowl's (1983) empirical interpretations of the geoid high associated with the African hotspot province being a fossil relict of the effects of the Pangean supercontinent on the lower mantle temperature field.

(*b*) *Dynamic topography*

The dynamic topography predicted at Earth's surface for our preferred model has a peak to peak amplitude of about 1 km, with uplifted regions generally associated with long-wavelength geoid highs (see Hager & Clayton 1988 for figures). Although there is no global data base of dynamically maintained topography for comparison, in regions where it is available, there is relatively good agreement between the predicted dynamic topography for our mantle-wide flow models and residual bathymetry not explained by plate cooling (Crough 1983; Cazenave *et al.* 1988). There is no such agreement for the flow models with a change in compositions at 670 km depth. The predicted dynamic topography is nearly independent of the present locations of continents and oceans. If these patterns are fixed relative to a high-viscosity lower mantle, relative motions of the continents and ocean basins should lead to changes of eustatic sea level.

Observations of the effects of core–mantle coupling on nutation amplitudes (Gwinn *et al.* 1986) and changes of length of day (Hide 1986) suggest that the topography of the CMB has an amplitude of hundreds of metres. It was to suppress the CMB topography calculated for our earlier models (Hager *et al.* 1985; Hager & Clayton 1988) that we included D″ in the models presented here. We have also argued that a low viscosity, perhaps stratified, D″ is to be expected *a priori*. The 10 km of relief at the CMB reported by Morelli & Dziewonski (1987) seem at first inconsistent with our expectations for much smaller amplitude dynamically maintained topography there.

There are at least two possible resolutions to the apparent paradox. The first is that substantial topography would be expected on the top of a chemically distinct D″. Equation (6) indicates that the dynamic topography is inversely proportional to the density contrast across a boundary. Because the density contrast between D″ and the overlying mantle is much smaller than the density contrast across the CMB, the top of a chemically distinct D″ would be expected to have very large (*ca.* 100 km) dynamic topography. The tradeoff between D″ velocity structure and thickness and CMB deflection should be considered for seismological inferences of CMB topography (Gudmundsson *et al.* 1987).

Alternatively, if there were a layer of molten silicate between the solid mantle and molten iron core, substantial dynamic topography of the solid–molten silicate boundary would result. For a given dynamic stress, boundary deflection is inversely proportional to the density contrast across a boundary (equation (6)). A layer of molten silicate at the top of the core would have a much smaller density contrast with the mantle than molten iron has, allowing much larger deflection of the boundary between solid mantle and fluid 'core'. (It is this high-impedance contrast boundary that would be seen by seismic reflection studies.) The boundary between molten silicate and molten metal would be difficult to detect using seismology, so the existence of such a layer would be difficult to detect directly. Its effects on nutation have not yet been worked out. Filling in of upwarpings of the solid boundary by 'anti-oceans' of molten silicate would shield the solid mantle from mechanical interaction with the molten iron core, allowing the change in length of day constraints to be satisfied.

5. Conclusions

Our ability to successfully model the observed geoid by using fluid-mechanical models with input of density fields inferred from seismic tomography gives us confidence both in the tomographic inversions, at least as smoothed by the long-wavelength geoid kernels, and in the importance of dynamic topography. We can fit the geoid either with models with a chemical barrier to flow at the 670 km seismic discontinuity or with models with a substantial viscosity increase with depth. The lack of observed large topography at this discontinuity leads us to prefer the latter model, i.e. deep-mantle convection.

A high-viscosity lower mantle would lead to a stratification of the style of mantle convection, without totally separating the upper and lower mantle. The slow mixing and long residence times inferred for such a high-viscosity lower mantle are consistent with a number of geochemical and geophysical observations.

Although our models are fairly successful, they are based upon a number of simplifying assumptions that should be removed in future work. Most notably, the assumption of spherical symmetry of the effective viscosity should be removed, particularly in the lithosphere. This will require substantial computing resources, but there are indications of systematic effects in the residual geoid (figure 6) that suggest that lateral variations are somewhat important. For example, residual geoid highs are associated with groups of hotspots in the central Indian Ocean, the North Atlantic, the Southeast Pacific, and the Basin and Range Province, perhaps indicative of locally negative geoid kernels resulting from lateral variations in viscosity in these regions.

Better constrained slab locations are becoming available through seismic tomography (Zhou & Clayton 1987). These can be used to improve the *ad hoc* slab model used. Considerations of solid-state theory may allow more realistic parametrizations of variations of the ratios P and S with depth. Self compression should also be included.

Finally, although the models presented here were computed by using trial-and-error forward modelling, we probably now understand the parametrization of the problem well enough to attempt a nonlinear inversion. This approach should enable us to explain more of the variance in the observed geoid, as well as to refine our models of mantle viscosity structure.

Close interaction with R. W. Clayton, R. P. Comer, and R. J. O'Connell was essential to this research. Critical reviews by D. L. Anderson, U. R. Christensen, W. Kiefer, and B. Parsons improved the manuscript substantially. This work was supported by NASA grants NAG5-315 and NAG5-842. This is contribution number 4625, Division of Geological and Planetary Sciences, California Institute of Technology, Pasadena, California 91125.

References

Anderson, D. L. 1982 *Nature, Lond.* **297**, 391–393.
Anderson, D. L. 1987 *J. geophys. Res.* **92**, 13968–13980.
Ashby, M. F. & Verrall, R. A. 1977 *Phil. Trans. R. Soc. Lond.* A **288**, 59–95.
Cazenave, A., Dominh, K., Rabinowicz, M. & Ceuleneer, G. 1988 *J. geophys. Res.* **93**, 8064–8077.
Chase, C. G. 1979 *Nature, Lond.* **282**, 464–468.
Chase, C. G. & McNutt, M. K. 1982 *Geophys. Res. Lett.* **9**, 29–32.
Chase, C. G. & Sprowl, D. R. 1983 *Earth planet. Sci. Lett.* **62**, 314–320.
Christensen, U. R. & Yuen, D. A. 1984 *J. geophys. Res.* **89**, 4389–4402.
Clayton, R. W. & Comer, R. P. 1984 *Terra Cognita* **4**, 282–283.
Creager, K. C. & Jordan, T. H. 1984 *J. geophys. Res.* **89**, 3031–3049.
Creager, K. C. & Jordan, T. H. 1986 *J. geophys. Res.* **91**, 3573–3589.
Crough, S. T. 1983 *A. Rev. Earth planet. Sci.* **11**, 165–193.
Crough, S. T. & Jurdy, D. M. 1980 *Earth planet. Sci. Lett.* **48**, 15–22.
Douglas, B. C., McAdoo, D. C. & Cheney, R. E. 1987 *Rev. Geophys.* **25**, 875–880.
Dziewonski, A. M. 1984 *J. geophys. Res.* **89**, 5929–5952.
Dziewonski, A. M. & Anderson, D. L. 1981 *Physics Earth planet. Int.* **25**, 297–356.

Dziewonski, A. M., Hager, B. H. & O'Connell, R. J. 1977 *J. geophys. Res.* **82**, 239–255.
Gantmacher, F. R. 1960 *The theory of matrices*, vols 1 and 2 (transl. from Russian by K. A. Hirsch). New York: Chelsea.
Grand, S. P. 1987 *J. geophys. Res.* **92**, 14065–14090.
Gudmundsson, O., Clayton, R. W. & Anderson, D. L. 1987 *Eos, Wash.* **68**, 1378.
Gurnis, M. & Hager, B. H. 1988 *Nature, Lond.* **335**, 317–321.
Gwinn, C. R., Herring, T. A. & Shapiro, I. I. 1986 *J. geophys. Res.* **91**, 4755–4765.
Hager, B. H. 1983 *Earth planet. Sci. Lett.* **63**, 97–109.
Hager, B. H. 1984 *J. geophys. Res.* **89**, 6003–6015.
Hager, B. H. & Clayton, R. W. 1988 In *Mantle convection* (ed. W. R. Peltier), pp. 657–763. New York: Gordon and Breach.
Hager, B. H., Clayton, R. W., Richards, M. A., Comer, R. P. & Dziewonski, A. M. 1985 *Nature, Lond.* **313**, 541–545.
Hager, B. H. & O'Connell, R. J. 1981 *J. geophys. Res.* **86**, 4843–4867.
Hager, B. H. & Raefsky, A. 1981 *Eos, Wash.* **62**, 1074.
Hide, R. 1986 *Q. Jl R. astr. Soc.* **27**, 30–20.
Isacks, B. & Molnar, P. 1971 *Rev. Geophys. space Phys.* **9**, 103–174.
Jordan, T. H. 1977 *J. Geophys.* **43**, 473–496.
Kaula, W. M. 1972 *The nature of the solid Earth* (ed. E. C. Robertson), pp. 386–405. New York: McGraw-Hill.
Kaula, W. M. 1980 *J. geophys. Res.* **85**, 7031–7044.
Lay, T. & Helmberger, D. V. 1983*a* *J. geophys. Res.* **88**, 8160–8170.
Lay, T. & Helmberger, D. V. 1983*b* *Geophys. Jl R. astr. Soc.* **75**, 799–837.
Lerch, F. J., Klosko, S. M. & Patch, G. B. 1983 NASA Tech. Memo 84986.
Masters, G., Jordan, T. H., Silver, P. G. & Gilbert, F. 1982 *Nature, Lond.* **298**, 609–613.
Morelli, A. & Dziewonski, A. M. 1987 *Nature, Lond.* **325**, 678–683.
Nakiboglu, S. M. 1982 *Phys. Earth planet. Int.* **28**, 302–311.
Nataf, H.-C., Nakanishi, I. & Anderson, D. L. 1984 *Geophys. Res. Lett.* **11**, 109–112.
Nataf, H.-C., Nakanishi, I. & Anderson, D. L. 1986 *J. geophys. Res.* **91**, 7261–7307.
O'Connell, R. J. & Hager, B. H. 1984 *Eos, Wash.* **65**, 1093.
O'Connell, R. J., Hager, B. H. & Richards, M. A. 1984 *Terra Cognita* **4**, 261.
Pekeris, C. L. 1935 *Mon. Not. R. astron. Soc. geophys. Suppl.* **3**, 343–367.
Peltier, W. R. & Wu, P. 1982 *Geophys. Res. Lett.* **9**, 731–734.
Revenaugh, J. & Parsons, B. 1987 *Geophys. Jl R. astro. Soc.* **90**, 349–368.
Ricard, Y., Fleitout, L. & Froidevaux, C. 1984 *Anns Geophysicae* **2**, 267–286.
Richards, M. A. & Wicks, C. W. 1987 *Eos, Wash.* **68**, 1379.
Richards, M. A. & Hager, B. H. 1984 *J. geophys. Res.* **89**, 5987–6002.
Richards, M. A. & Hager, B. H. 1988*a* *The physics of planets* (ed. S. K. Runcorn), pp. 247–277. Chichester: John Wiley.
Richards, M. A. & Hager, B. H. 1988*b* *J. geophys. Res.* (In the press.)
Richards, M. A., Hager, B. H. & Sleep, N. H. 1988 *J. geophys. Res.* **93**, 7690–7708.
Runcorn, S. K. 1964 *J. geophys. Res.* **69**, 4389–4394.
Runcorn, S. K. 1967 *Geophys. Jl R. astr. Soc.* **42**, 375–384.
Sammis, C. G., Smith, J. C., Schubert, G. & Yuen, D. A. 1977 *J. geophys. Res.* **82**, 3747–3761.
Schubert, G., Yuen, D. A. & Turcotte, D. L. 1975 *Geophys. Jl R. astr. Soc.* **42**, 705–735.
Tanimoto, T. 1986 *Geophys. Jl R. astr. Soc.* **84**, 49–69.
Tanimoto, T. & Anderson, D. L. 1984 *Geophys. Res. Lett.* **4**, 287–290.
Vassiliou, M. S., Hager, B. H. & Raefsky, A. 1984 *J. Geodyn.* **1**, 11–28.
Woodhouse, J. H. & Dziewonski, A. M. 1984 *J. geophys. Res.* **89**, 5953–5986.
Young, C. J. & Lay, T. 1987 *A. Rev. Earth planet. Sci.* **15**, 25–46.
Zhang, S. & Yuen, D. A. 1987 *Geophys. Res. Lett.* **14**, 899–902.
Zhou, H. W. & Clayton, R. W. 1987 *Eos, Wash.* **68**, 1379.

Phil. Trans. R. Soc. Lond. A **328**, 329–349 (1989)
Printed in Great Britain

Low-frequency seismology and the three-dimensional structure of the Earth

By T. G. Masters

Institute of Geophysics and Planetary Physics, University of California, San Diego, A-025, La Jolla, California 92093, *U.S.A.*

We attempt to catalogue those features of the three-dimensional structure of the Earth that are well-constrained by low-frequency data (i.e. periods longer than about 125 seconds). The dominant signals in such data are the surface-wave equivalent modes whose phase characteristics are mainly affected by a large scale structure of harmonic degree two in the upper mantle. Available aspherical models predict this phase behaviour quite well, but do not give an accurate prediction of the observed waveforms and we must appeal to higher-order structure and/or coupling effects to give the observed complexity of the data. Strong splitting of modes which sample the cores of the Earth is also observed and, though we do not yet have a model of aspherical structure which gives quantitative agreement with these data, anisotropy or large-scale aspherical structure in the inner core appears to be required to model the observed signal.

Introduction

The last few years have seen some major advances in our understanding of the large-scale aspherical structure of the Earth and it is the intention of this paper to summarize the contributions of low-frequency seismology. There are several advantages to working with low-frequency data. These are the only seismic data that are at all sensitive to the density structure of the Earth (though it will be some time before aspherical density structure is reliably determined). Furthermore, relatively simple representations of the source suffice to model the excitation of long period waves and we can be reasonably sure (for all but the largest earthquakes) that any anomalous signal we see is due to structure. Finally, attenuation appears to be only weakly frequency dependent in the free oscillation frequency band, so simplifying the modelling process. On the other hand, there are some disadvantages associated with long-period data. Perhaps the most severe of these is the low sensitivity to structure of odd harmonic degree. As we shall see, the features in the data which constrain the odd-order structure are quite subtle and we must be careful that any theoretical approximations that we make do not swamp this signal with spurious signal-generated noise. An obviously desirable way to proceed is to use both low-frequency data and shorter-period body wave data to constrain structure in a simultaneous inversion and this will undoubtedly become common in the near future.

In low-frequency seismology, it is convenient to specify structure by expansion in spherical harmonics and several models of aspherical structure incorporating structure up to harmonic degree eight or higher now exist. An alternative representation of structure calls for an *a priori* division of the surface of the sphere into various regions which are supposed to have similar seismic properties. Such *tectonic regionalizations* are, of course, subject to the prejudice of the individual investigator and may not have the correct degrees of freedom required to model the data. On the other hand, they allow the specification of sharp lateral changes in structure with relatively few unknown model parameters while an equivalent spherical harmonic expansion extends to very high harmonic degree. In practice, we perform calculations of synthetic seismograms by using the spherical harmonic basis because coupling between free oscillations of varying harmonic degree depends only upon some of the harmonic components of the model. Thus, models specified as a tectonic regionalization are first expanded in spherical harmonics.

One of the topics we consider in this paper is the qualitative effect of short wavelength structure on low-frequency data by computing synthetic seismograms for a tectonic regionalization expanded up to harmonic degree 40. The reason for our interest in such calculations is that the existing models leave much of the signal in the data unexplained. There are many possible reasons for this, e.g. approximate theories relating the data to the model have been used in the analyses and smooth models with only a few degrees of freedom have been sought. It is probable that some features of the data (e.g. surface wave amplitude anomalies) will require relatively short wavelength structure to model them and great care must be taken that any theoretical approximations will be adequate to the task. We are fortunate in low-frequency seismology that there are relatively complete algorithms for computing synthetic seismograms on an aspherical Earth (see, for example, Park & Gilbert 1986). Thus, given a model of three-dimensional structure, we are able to compute accurate synthetic seismograms for comparison with the data. This is still a computationally heavy task so an important aspect of such calculations is that they allow us to check approximate methods of seismogram calculation.

In the next section, we briefly consider the theoretical basis for the calculation of synthetic seismograms and show how differential seismograms can be constructed to allow fitting of complete waveforms or the spectra of small groups of modes. It is an observational fact that weakly coupled, split multiplets usually look like single resonance functions in the frequency domain so we consider an approximate theory which allows us to model the apparent centre frequencies and attenuation rates of such modes. Following sections discuss the ability of some current types of aspherical models to fit the observations and we make an attempt to catalogue those features of the Earth's aspherical structure which appear to have been robustly determined from low-frequency data.

Theoretical background

Low-frequency seismic data are most naturally analysed in the frequency domain and it is convenient to first consider the effect of aspherical structure on *isolated* multiplets. In reality, it is probably true that no multiplet can truly be regarded as isolated though coupling between multiplets is usually weak and most of the anomalous signal due to aspherical structure is caused by the interaction of singlets within a single multiplet. Coupling between multiplets causes small but measurable additional signals. First, consider the seismogram on a *spherical* Earth:

$$\boldsymbol{s}(\boldsymbol{r}, t) = \sum_{k} \sum_{m=-l}^{l} \boldsymbol{\sigma}_k^m(\boldsymbol{r})\, a_k^m(\boldsymbol{r}_0)\, \mathrm{e}^{\mathrm{i}\omega_k t}, \tag{1}$$

where the real part is understood, $\boldsymbol{\sigma}_k^m$ are the $2l+1$ singlet eigenfunctions of the kth multiplet given by

$$\boldsymbol{\sigma}_k^m(\boldsymbol{r}) = \hat{\boldsymbol{r}} U_k(\boldsymbol{r}) Y_l^m(\theta, \phi) + V_k(r) \nabla_1 Y_l^m(\theta, \phi) - W_k(r) \hat{\boldsymbol{r}} \times \nabla_1 Y_l^m(\theta, \phi), \tag{2}$$

where $\nabla_1 = \hat{\theta}\partial_\theta + \operatorname{cosec}(\theta)\,\hat{\phi}\partial_\phi$. For a point source at r_0 with moment rate tensor M, the excitation coefficients, a_k^m, are given by $a_k^m = M:\epsilon_k^{m*}(r_0)$, where ϵ_k^{m*} is the complex conjugate of the strain tensor of the mth singlet (Gilbert & Dziewonski 1975); ω_k is the degenerate frequency of the multiplet.

For notational convenience, we can consider σ_k^m and a_k^m to be arrays $2l+1$ in length so that the sum over m in (1) can be rewritten as a dot product. No confusion as to the vector nature of σ will arise if we consider only a single component of recording, i.e.

$$s(\mathbf{r},t) = \sum_k \boldsymbol{\sigma}_k(\mathbf{r}) \cdot \mathbf{a}_k(\mathbf{r}_0)\, \mathrm{e}^{\mathrm{i}\omega_k t}. \tag{3}$$

The singlet eigenfunctions satisfy a variational principle, which we write schematically as

$$0 = L = V_0(\boldsymbol{\sigma}_i^*, \boldsymbol{\sigma}_j) - \omega_k^2\, T_0(\boldsymbol{\sigma}_i^*, \boldsymbol{\sigma}_j), \tag{4}$$

with $\delta L = 0$. V_0 is related to the potential energy and $\omega_k^2 T_0$ is related to the kinetic energy. T_0 is given by

$$T_0 = \int_V \rho_0 \boldsymbol{\sigma}_i^* \cdot \boldsymbol{\sigma}_j \,\mathrm{d}V, \tag{5}$$

where ρ_0 is the density in the spherically averaged model. The eigenfunctions are orthogonal and are usually normalized so that

$$T_{0ij} = \delta_{ij}. \tag{6}$$

In the presence of rotation and small perturbations in structure, the symmetry of the original model is destroyed and the degeneracy is removed. Each of the $2l+1$ singlets in equation (1) now has a slightly different frequency. The variational principle now reads:

$$0 = L = V(\mathbf{s}^*, \mathbf{s}) + \omega W(\mathbf{s}^*, \mathbf{s}) - \omega^2 T(\mathbf{s}^*, \mathbf{s}), \tag{7}$$

with $\delta L = 0$. W is the Coriolis force interaction functional and V and T are the potential energy and kinetic energy interaction functionals (explicit expressions for these functionals can be found in Part & Gilbert 1986). We now expand $\mathbf{s}$ as a linear combination of spherical Earth singlets so that the new singlets are given by

$$\mathbf{s}_j = \sum_m U_{mj}\, \boldsymbol{\sigma}_k^m. \tag{8}$$

Substitution into (7) leads to a quadratic eigenvalue problem, but it is usually sufficiently accurate to replace ω in the Coriolis term with an average frequency, ω_k, say, so that equation (7) reduces to an ordinary eigenvalue problem (Park & Gilbert 1986). With the normalization given above, we have

$$V(\boldsymbol{\sigma}_i, \boldsymbol{\sigma}_j) = V_{ij} = \omega_k^2\, \delta_{ij} + \delta V_{ij},$$

$$T(\boldsymbol{\sigma}_i, \boldsymbol{\sigma}_j) = T_{ij} = \delta_{ij} + \delta T_{ij},$$

and can then define the *splitting matrix* $\mathbf{H}$

$$2\omega_k \mathbf{H} = \delta \mathbf{V} + \omega_k \mathbf{W} - \omega_k^2\, \delta \mathbf{T}. \tag{9}$$

With these definitions, we end up solving the following eigenvalue problem for the kth multiplet:

$$\mathbf{H}\mathbf{U} = \mathbf{U}\boldsymbol{\Omega}, \tag{10}$$

where $\omega_k + \Omega_{jj}$ are the new singlet frequencies and the eigenfunctions of $\boldsymbol{H}$ allow us to construct the new singlet eigenfunctions by using equation (8). The elements of the splitting matrix are given by (Woodhouse & Dahlen 1978)

$$H_{mm'} = \omega_k(a_k + mb_k + m^2 c_k)\,\delta_{mm'} + \sum_{s=2}^{2l} \gamma_s^{mm'}\, c_s^{m-m'}. \tag{11}$$

The first term gives the contribution of rotation and hydrostatic ellipticity of figure while the second gives the contribution of all other structure as specified by its spherical harmonic expansion, i.e. we let

$$\delta \boldsymbol{m}(\boldsymbol{r}) = \sum_{s,t} \delta \boldsymbol{m}_s^t(r)\, Y_s^t(\theta, \phi), \tag{12}$$

where

$$\delta \boldsymbol{m}_s^t(r) = (\delta\rho_s^t(r), \delta\kappa_s^t(r), \delta\mu_s^t(r), h_{sj}^t, \dots), \tag{13}$$

with ρ being density, κ bulk modulus, h_j the radius of the jth discontinuity, etc. Then c_s^t is a linear functional of the s, t coefficient of each model parameter i.e.

$$c_s^t = \int_0^a \delta \boldsymbol{m}_s^t(r) \cdot \boldsymbol{G}_s(r)\, r^2\, \mathrm{d}r \tag{14}$$

and expressions for the kernels, $\boldsymbol{G}$, can be found in Woodhouse & Dahlen (1978). Finally, note that $\gamma_s^{mm'}$ can be written in terms of Wigner $3j$ symbols which can be computed using the recurrence relations given by Schulten & Gordon (1975). The $3j$ symbols are zero unless certain *selection rules* are satisfied. For an isolated multiplet, there is no contribution to (11) unless $0 \leqslant s \leqslant 2l$, s is even and $t = m - m'$.

The main point that we wish to emphasize is that the effect of aspherical structure on a particular multiplet is completely specified by the c_s^t which henceforth we shall call *structure coefficients*. Of course, we cannot directly determine the splitting matrix (and hence the structure coefficients) from the data because the seismogram is a nonlinear function of the splitting matrix. To see this, we use the eigenvalue–eigenvector decomposition of $\boldsymbol{H}$ given by equation (10) so that the aspherical Earth equivalent of equation (3) is

$$s(\boldsymbol{r}, t) = \sum_k (\boldsymbol{\sigma}_k\, \boldsymbol{U}_k)\, \mathrm{e}^{\mathrm{i}(\boldsymbol{\Omega}_k + \omega_k)t}\, (\boldsymbol{U}_k^{-1}\, \boldsymbol{a}_k). \tag{15}$$

This equation shows how each singlet on the aspherical Earth is, in general, a linear combination of all the singlets of the spherical Earth. An interesting exception arises if the multiplet is dominantly sensitive to axisymmetric structure. Such structure contributes only to the diagonal elements of $\boldsymbol{H}_k$ so $\boldsymbol{U}_k$ is, in this case, the unit matrix and there is no mixing of the spherical Earth singlets.

Equation (15) can be written in several ways which make the effect of aspherical structure clearer (Woodhouse & Girnius 1982). For example, we can regard $\boldsymbol{a}_k$ as being a slowly varying function of time which satisfies the equation:

$$\mathrm{d}\boldsymbol{a}_k/\mathrm{d}t = \mathrm{i}\boldsymbol{H}_k\, \boldsymbol{a}_k(t), \tag{16}$$

with the initial condition that at $t = 0$, $\boldsymbol{a}_k$ is defined as in equation (1). Then

$$s(\boldsymbol{r}, t) = \sum_k \boldsymbol{\sigma}_k(\boldsymbol{r}) \cdot \boldsymbol{a}_k(\boldsymbol{r}_0, t)\, \mathrm{e}^{\mathrm{i}\omega_k t}. \tag{17}$$

Equation (16) has the solution

$$\boldsymbol{a}_k(t) = \boldsymbol{P}_k(t)\, \boldsymbol{a}_k(0), \tag{18}$$

where $\boldsymbol{P}_k(t) = \exp(\mathrm{i}\boldsymbol{H}_k t)$ is the matrizant or propagator matrix; $\boldsymbol{a}_k$ can be thought of as an envelope function which varies on a long timescale because the magnitude of the splitting matrix is of the same order as the splitting width of the multiplet. Note that aspherical structure only appears in equation (17) through the time-dependence of $\boldsymbol{a}_k$.

We can extend this representation to compute differential seismograms so that the effect of perturbing a structure coefficient on the time series can be calculated. To do this, we differentiate equation (17) with respect to one of the c_s^t giving

$$\frac{\partial s}{\partial c_s^t} = \sum_k \boldsymbol{\sigma}_k \cdot \frac{\partial \boldsymbol{a}_k}{\partial c_s^t} \mathrm{e}^{\mathrm{i}\omega_k t}, \tag{19}$$

where $\partial \boldsymbol{a}_k / \partial c_s^t$ can be evaluated by differentiating equation (16) with respect to the c_s^t so yielding an inhomogeneous propagator equation whose solution is well known (Gilbert & Backus 1966; Ritzwoller *et al.* 1986). The differential seismograms may now be used to iteratively improve a set of structure coefficients so that equation (17) gives a good representation of the data (Woodhouse & Giardini 1985; Ritzwoller *et al.* 1986, 1988).

Before we leave the case of an isolated multiplet, we recast equation (17) into yet another form which is useful in developing an approximate theory to describe the spectra of unresolvably split multiplets. If we define a time dependent *location parameter* for the kth multiplet as

$$\lambda(t) = \boldsymbol{\sigma} \cdot \boldsymbol{H} \cdot \boldsymbol{a}(t) / \boldsymbol{\sigma} \cdot \boldsymbol{a}(t), \tag{20}$$

then equation (17) can be written

$$s(\boldsymbol{r}, t) = (\boldsymbol{\sigma} \cdot \boldsymbol{a}(0)) \exp\left(\mathrm{i} \int_0^t \lambda(t)\, \mathrm{d}t\right) \exp(\mathrm{i}\omega_k t). \tag{21}$$

If $\lambda(t)$ is only weakly dependent upon time, i.e. $\lambda(t) \approx \lambda(0) = \lambda_0$ say, then equation (21) reduces to

$$s(\boldsymbol{r}, t) = (\boldsymbol{\sigma} \cdot \boldsymbol{a}(0)) \exp(\mathrm{i}(\omega_k + \lambda_0)\, t), \tag{22}$$

which corresponds to a peak shift in the spectrum. This is the asymptotic result of Jordan (1978) and Dahlen (1979), which is valid in the limit that the wavelength of the structure is much longer than the wavelength of the mode (i.e. $s \ll l$). This result appears to be roughly valid for surface-wave equivalent modes with harmonic degree greater than about 20 for model M84A of Woodhouse & Dziewonski (1984). Figure 1 shows the actual time variation of $\lambda(t)$ for some modes and we do a much better job of modelling the data if we replace λ_0 in equation (23) with a time-average of the true $\lambda(t)$. Smith & Masters (1989) show how suitable time averages can be constructed which give a *complex* peak shift so modifying the apparent attenuation rate as well as the centre frequency of an unresolvably split multiplet. This theory works well with synthetic data constructed using M84A as the test model, but does not do a very good job of explaining observed peak shifts. Two possible reasons suggest themselves: (1) the Earth has significant power in structure of higher harmonic degree than is present in M84A and (2) coupling between multiplets significantly affects the observed peak shifts. This latter hypothesis can be tested by computing coupled mode synthetic seismograms.

Coupling between multiplets can be accommodated by returning to equation (7) and expanding $\boldsymbol{s}$ in the singlets of all the coupling multiplets. Much of the theory described above can be extended to the coupled multiplet case in an obvious way. Even equation (21) can be extended to the weakly coupled case because the singlets of weakly coupled multiplets clump into groups forming *hybrid* multiplets. Each singlet of the hybrid multiplet is a linear

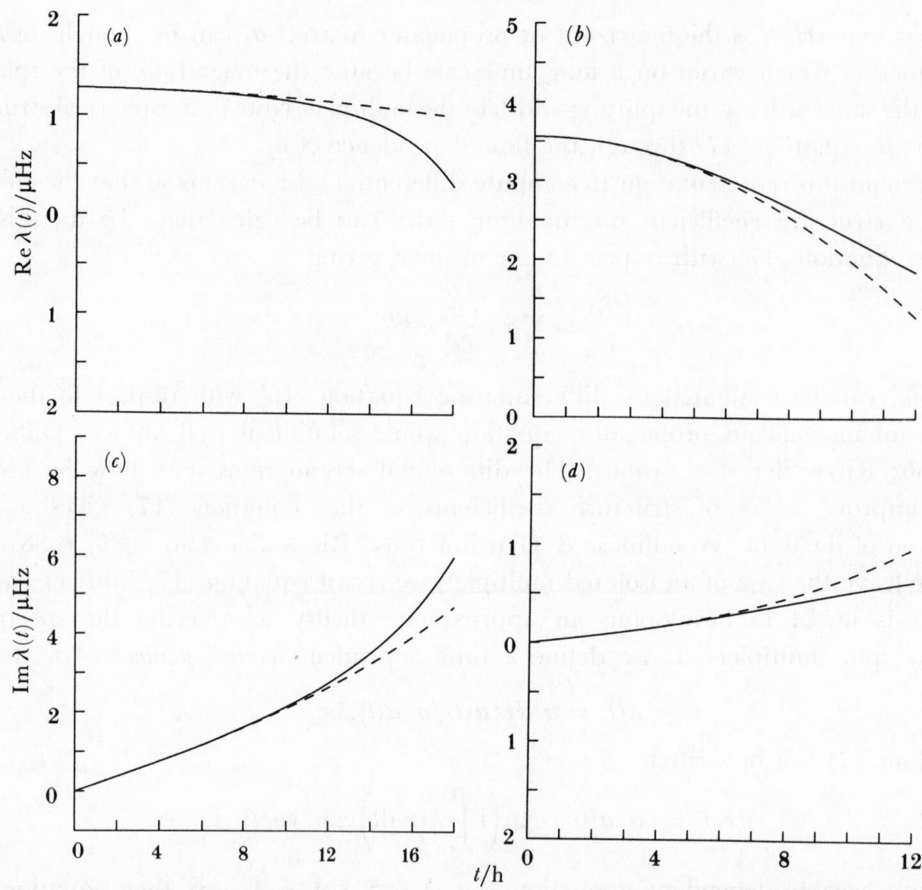

FIGURE 1. The generalized location parameter $\lambda(t)$ as a function of time. (*a*) and (*c*) show the real and imaginary parts of $\lambda(t)$ for ${}_0S_{24}$. (*b*) and (*d*) are the same but for ${}_0S_{36}$. The receiver was located in Guam for a source in Tonga. Model M84A with rotation was used to perform the calculation. Figure from Smith & Masters (1989).

combination of all the singlets of the coupling multiplets and it is possible to find the coefficients in this combination to first order in the coupling strength without performing a complete matrix decomposition (Park 1987; Dahlen 1987). Park calls this the *subspace projection method* and it may be used to define a location parameter for a hybrid multiplet. The selection rules for coupling between different multiplets can get quite complicated. The most obvious kind of coupling seen in the data is Coriolis coupling of ${}_nS_l$ modes to ${}_nT_{l\pm1}$ and in fact, significant coupling of fundamental spheroidal modes to fundamental toroidal modes occurs throughout the frequency band $1.5 \rightarrow 3$ mHz (Masters *et al.* 1983). Of course, this kind of coupling does not directly tell us about aspherical structure so of more interest is the fact that coupling of ${}_nS_l$ to its neighbours, ${}_nS_{l-1}$, ${}_nS_{l+1}$, etc., sensitizes the seismogram to structure of odd harmonic degree. We shall show some calculations which illustrate this effect later in the paper and an asymptotic treatment can be found in Park (1987) and Romanowicz (1987).

In the next sections we discuss some of the observations and the ability of available models to fit them. A more complete discussion of the low-frequency data can be found in Masters & Ritzwoller (1988). It is convenient to divide the observations into two categories; (1) large harmonic degree surface-wave equivalent modes which are dominantly sensitive to the structure of the upper mantle and (2) low harmonic degree, high Q modes which are also sensitive to the structure of the deep Earth.

Surface-wave modes and upper mantle models

The surface-wave equivalent modes are unresolvably split and may be amenable to interpretation by using asymptotic theory (assuming that the spectrum of aspherical structure is very red). The main approximation that is invoked in the asymptotic theory is that the data are sensitive to structure only along the great-circle path joining source and receiver. This approximation is theoretically justified for the mode peak shift data in the limit that $l \to \infty$ when equation (20) becomes

$$\lambda(t) = \lambda(0) = \sum_{s\,(\text{even})} P_s(0) \sum_t c_s^t\, Y_s^t(\Theta, \Phi), \tag{23}$$

where Θ, Φ is the location of the pole of the great circle joining the source and receiver (Jordan 1978). Centre frequencies and apparent attenuation rates are easy to measure for the highly excited fundamental spheroidal modes (figure 2) and, if one plots the observed peak shifts as symbols at the pole locations, one obtains maps such as in figure 3. A large-scale structure is

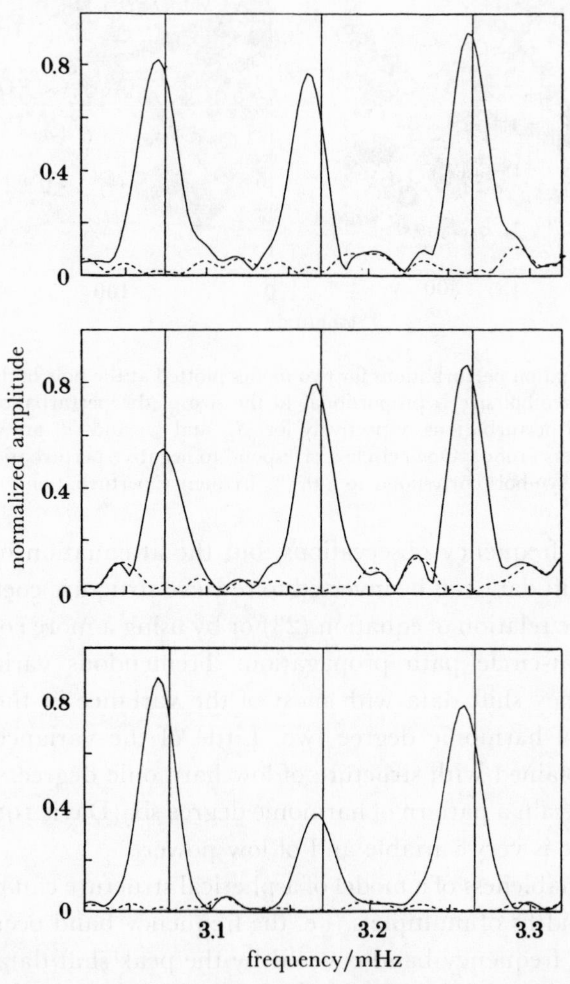

Figure 2. Linear amplitude spectra of a frequency band encompassing the modes ${}_0S_{22}$ to ${}_0S_{24}$. The solid line is the original spectrum while the dashed line is the residual after the best fitting resonance function has been removed from the data. Note that the resonance function model does a good job of modelling the data and that the apparent centre frequency of the modes is variable from record to record.

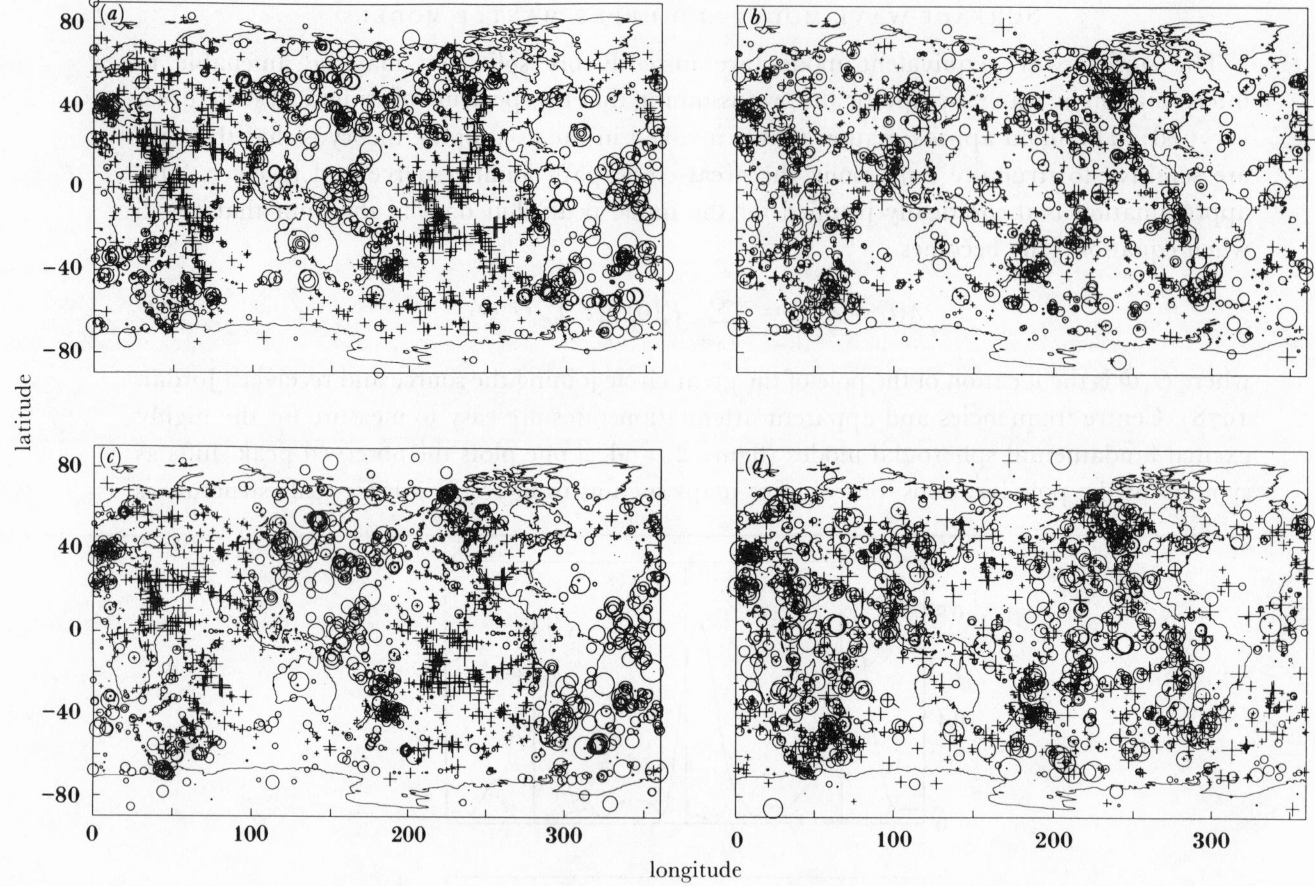

FIGURE 3. Frequency and attenuation perturbations for two modes plotted at the pole of the great-circle joining the source and receiver. The symbol size is proportional to the size of the perturbation. (*a*) and (*b*) show the frequency and attenuation perturbations respectively for ${}_0S_{22}$ and (*c*) and (*d*) are equivalent plots for ${}_0S_{42}$. Figure from Smith & Masters (1989). Open circles correspond to negative perturbations and pluses to positive perturbations. The largest symbols correspond to a 0.4% frequency perturbation.

apparent in the real part frequency observations, but the attenuation observations show no such pattern. The peak shift data can be inverted to retrieve structure coefficients (figure 4) by using either the asymptotic relation of equation (23) or by using a more complete theory which takes account of off-great-circle path propagation. Tremendous variance reductions are possible using the frequency shift data with most of the variance in the observations being explained by a pattern of harmonic degree two. Little of the variance in the attenuation measurements can be explained with structure of low harmonic degree. The frequency shifts also seem to robustly constrain a pattern of harmonic degree six (Davis 1987; Smith & Masters 1989), but other structure is very variable and of low power.

A gross test of the reasonableness of a model of aspherical structure can be had by comparing the theoretical splitting widths of multiplets (i.e. the frequency band occupied by the singlets of the multiplet) with the frequency band covered by the peak shift data (figure 5). We find that a model like M84A tends to underpredict the splitting widths of fundamental spheroidal modes but that a tectonic regionalization (see below) with a little more power in the lower

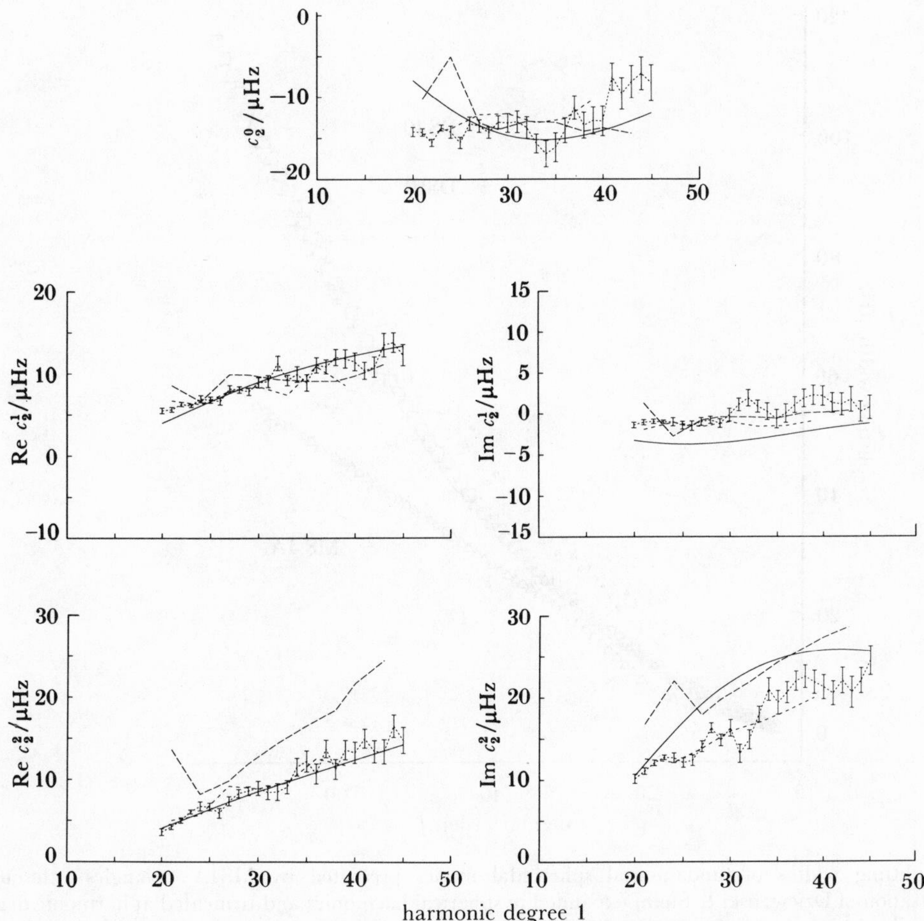

FIGURE 4. Structure coefficients, c_2^t for the fundamental spheroidal modes showing the coefficients of Smith & Masters (1989) (data with error bars) compared with the predictions of M84A (solid line), the results of Davis (1987) (intermediate dashed line), and the results of Nakanishi & Anderson (1983) (long dashed line). Agreement is quite good for these coefficients, but is worse at higher harmonic degrees. Figure from Smith & Masters (1989).

harmonic degrees is capable of giving a qualitative fit. Note that the regionalization including harmonics up to degree 40 gives almost identical results to the regionalization truncated at harmonic degree 10 so there is no *a priori* reason why models with sharp edges cannot give a good fit to the data.

Of course, one need not confine oneself to the interpretation of peak shift data and both waveform fitting techniques (Woodhouse & Dziewonski 1984; Tanimoto 1987) and surface wave dispersion measurements (Nakanishi & Anderson 1983, 1984) have been used. All interpretations have assumed that off-path propagation can be ignored. The waveform fitting experiment of Woodhouse & Dziewonski deserves further comment as these authors do not try to constrain the structure coefficients of individual multiplets, but directly invert for a parametrized form of $\delta \boldsymbol{m}_s^t(r)$ (see equation (14)). The predictions of their model M84A of the structure coefficients of some fundamental modes are compared with the results of the peak

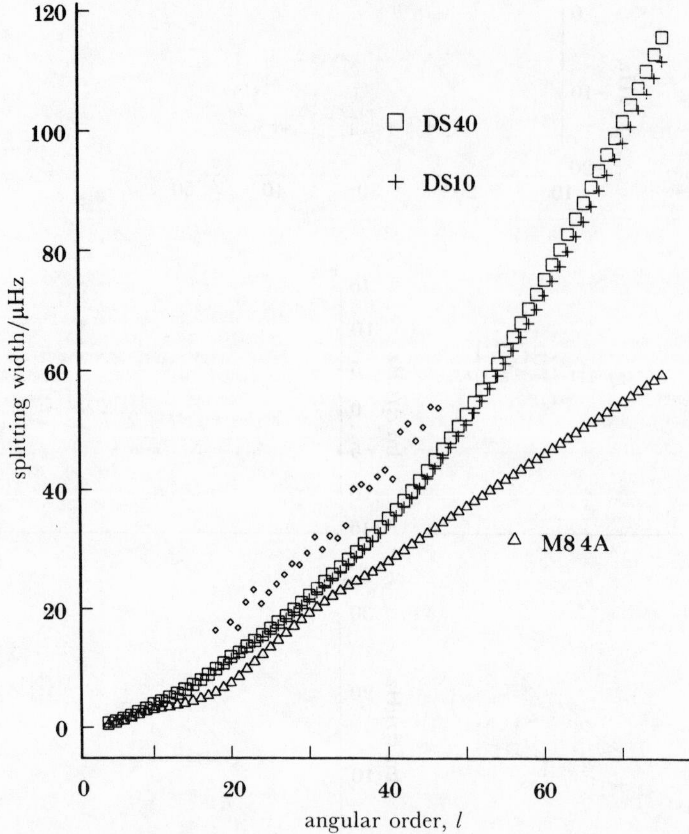

FIGURE 5. Splitting widths of fundamental spheroidal modes predicted by M84A (triangles), the tectonic regionalization of Dziewonski & Steim expanded in spherical harmonics and truncated at harmonic degree 40 (DS40) or harmonic degree 10 (DS10). The diamonds are the observations.

fitting experiment in figure 4. A convenient way of comparing different models is to plot the *splitting functions* (Woodhouse *et al.* 1986) defined by

$$\sum_{s,t} c_s^t \, Y_s^t(\theta, \phi). \tag{24}$$

We illustrate this in figure 6, where we see that agreement of the degree 2 pattern is quite good (as it should be as this accounts for over 60% of the variance in the data), but even the quite well-determined degree 6 structure shows significant differences. Numerical experiments indicate that the degree 6 structure is not reliably determined if the asymptotic theory (equation (23)) is used to interpret the peak shift data because the signal from non-asymptotic effects rivals that of the structure itself. Thus it may be true that the degree 6 structure in M84A is biased by the use of the great-circle approximation. As the power in harmonic degrees 4 and 8 is even smaller than that in degree 6, it is unsurprising that there is as yet little agreement as to the shapes of the splitting functions at these harmonic degrees. Comparisons with the surface-wave dispersion data show even less agreement which may reflect the difficulty of accurately determining phase velocity at very long periods.

Model M84A also includes structure of odd harmonic degree by using the difference in phase

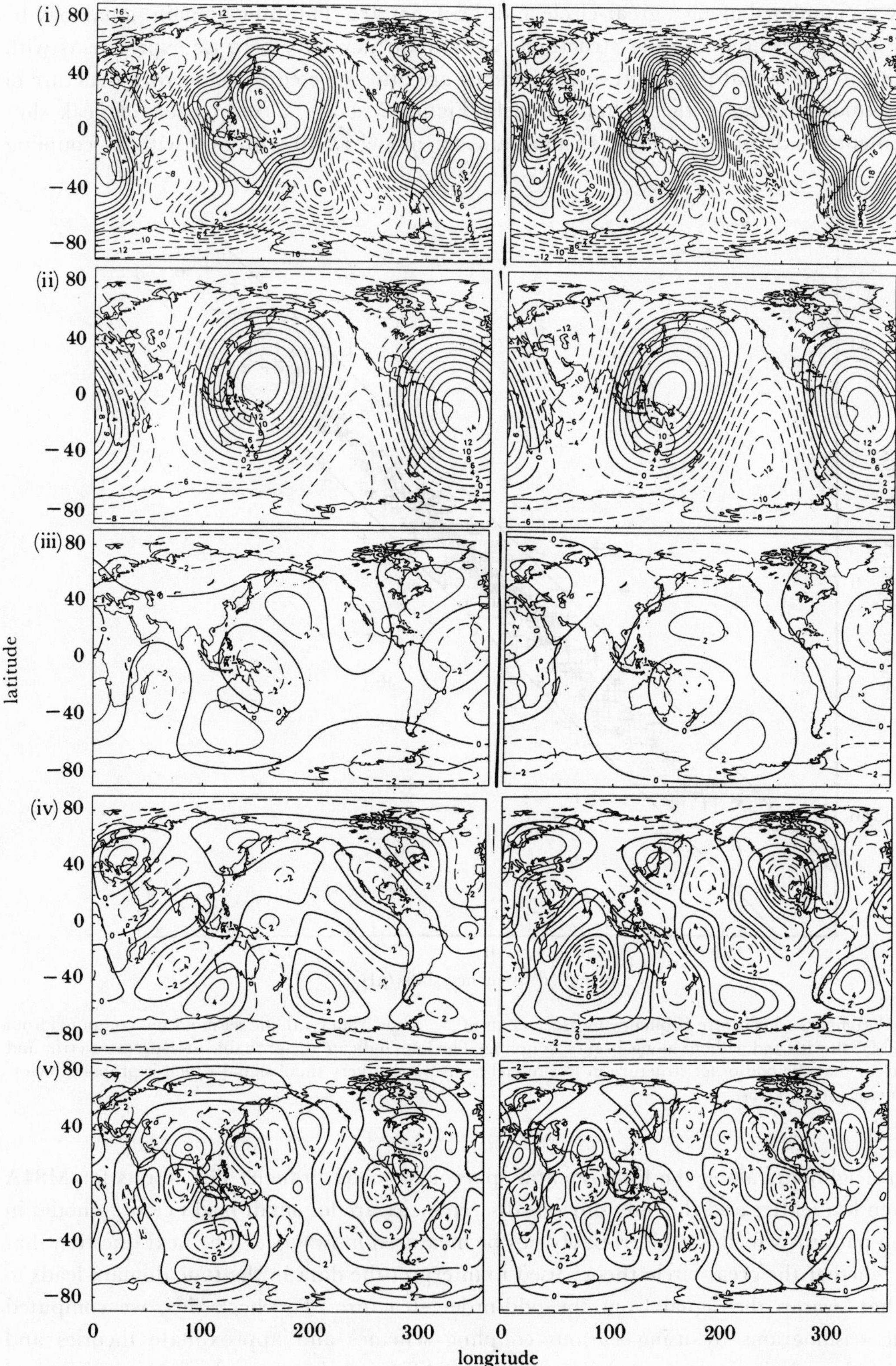

FIGURE 6. Contour plots of the splitting function of mode ${}_0S_{23}$ as determined from peak shift data (left column) and as predicted by M84A (right column); (i) degrees, 2–8, (ii) degree 2, (iii) degree 4, (iv) degree 6, (v) degree 8. Agreement is very good at degree 2 but deteriorates at higher harmonic degree. Note that degree 2 dominates the overall pattern. Figure from Smith & Masters (1989).

anomaly accumulated along a great-circle and the minor arc. This time domain signal may be the only way of seeing odd-order structure in low-frequency data because experiments with synthetic data show that peak shift measurements are little affected by odd-order structure of the magnitude present in the current models. Figure 7 shows a comparison of peak shift measurements made from synthetic seismograms for model DS40 with and without coupling

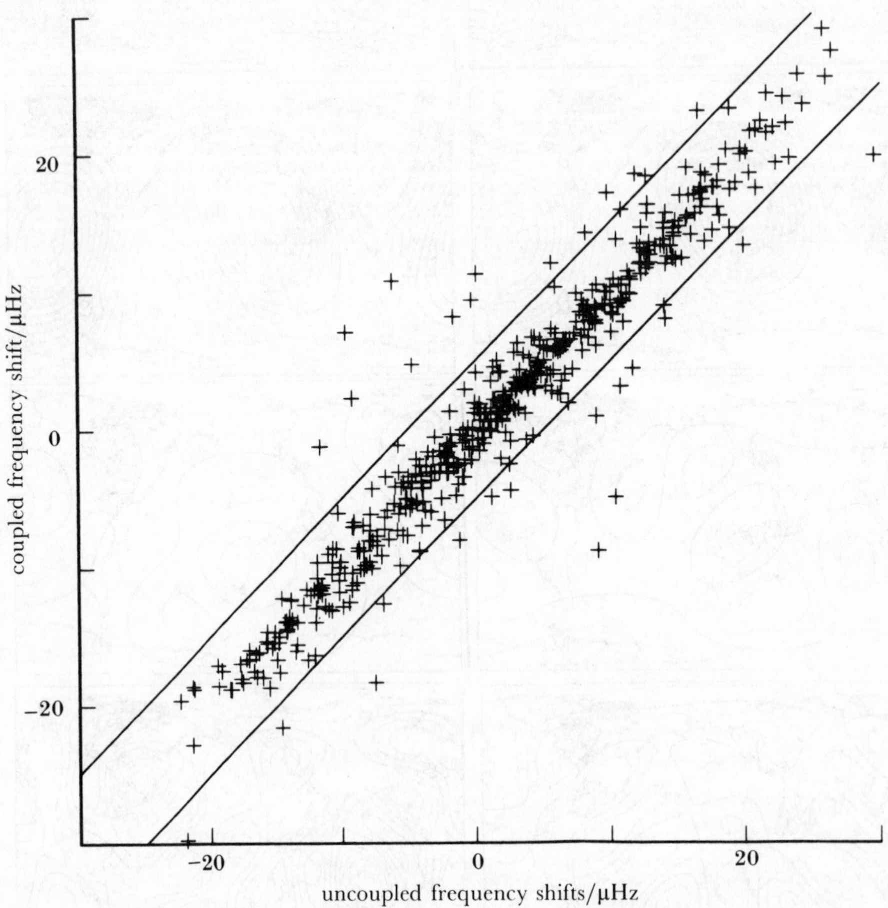

FIGURE 7. A comparison of centre frequency measurements of ${}_0S_{50}$ made from synthetic seismograms computed from model DS40 with and without along-branch coupling. The lines indicate the probable measurement error and it is clear that the odd-order structure in this model contributes a very small signal which is probably below the measurement error.

between neighbours along the fundamental spheroidal mode branch. The results for M84A show even less scatter and the only obvious discrepancies are for small peaks close to nodes in radiation patterns which would probably not be measured in practice. The next question that arises is whether the great-circle theory used to interpret the data in the time domain leads to errors that swamp the signal from the odd-order structure. To check this, we computed synthetic seismograms by using various coupling schemes and approximate theories and computed the relative root mean square (r.m.s.) difference between the first ten hours of recording and averaged the results over roughly 100 source-receiver pairs. We present the average results in matrix form in table 1. Seismograms constructed using the theory of

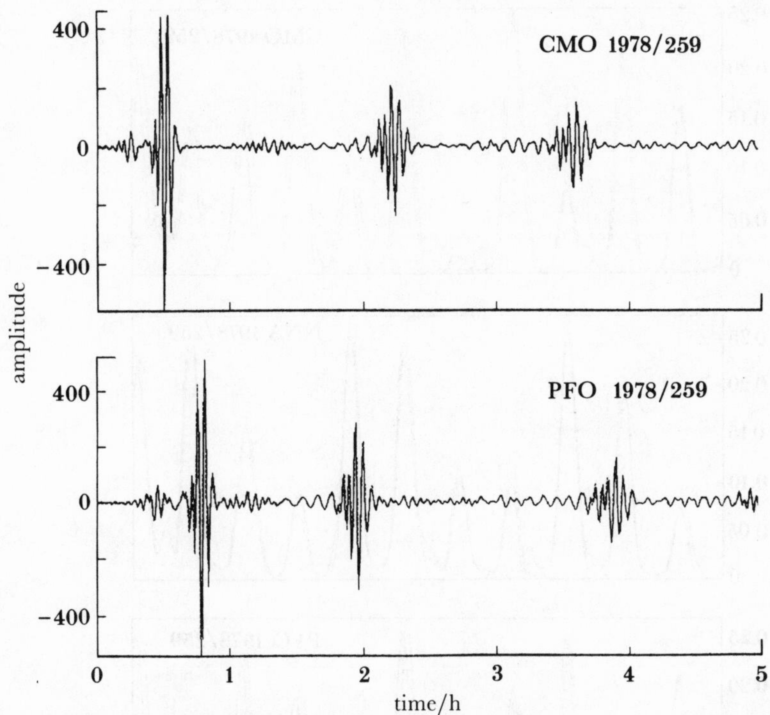

FIGURE 8. Comparisons of along-branch coupled mode synthetic seismograms (dashed line) with uncoupled mode synthetic seismograms (solid line). The case of CMO is typical while an extreme example is given by PFO. The calculations have been performed with M84A as the aspherical structure and include the first three branches up to a frequency of 8 mHz. The effect of coupling is so small that it is extremely difficult to distinguish the two calculations.

Woodhouse & Dziewonski do seem to fit the precise calculations slightly better than the calculations ignoring along-branch coupling (i.e. an r.m.s. difference of 13 % against 17 %). One thing that is noticeable from table 1 is that all the numbers are relatively small. An r.m.s. difference of about 15 % corresponds to an almost exact visual overlay of the time series (figures 8 and 9) and we would be well satisfied if we could fit the actual data to this precision. Needless

TABLE 1. MEAN RELATIVE R.M.S. RESIDUAL BETWEEN SEISMOGRAMS COMPUTED WITH DIFFERENT THEORETICAL SCHEMES

(The numbers are in percent and are the averages of the first 10 hours of 80 recordings. The aspherical model is M84A. 'Coupled-5' indicates a calculation where the singlets of five adjacent fundamental modes have been coupled together whereas 'coupled-3' implies a calculation with coupling to nearest neighbours. 'WD84' implies a calculation using a the theory of Woodhouse & Dziewonski (1984) while 'isolated' includes no mode coupling at all. 'Peak shift' implies a calculation with only the asymptotic shift in the peak frequencies and 'spherical' is the spherical Earth calculation. All fundamental spheroidal modes up to a frequency of 8 mHz have been included and the time series have been low-passed with a corner at 6.5 mHz to reduce the effect of ringing.)

	coupled-5	coupled-3	WD84	isolated	peak shift	spherical
coupled-5	0	9.8	13.0	18.3	19.6	30.5
coupled-3	—	0	13.6	17.0	18.3	29.6
WD84	—	—	0	20.7	18.7	31.1
isolated	—	—	—	0	5.9	22.0
peak shift	—	—	—	—	0	21.2
spherical	—	—	—	—	—	0

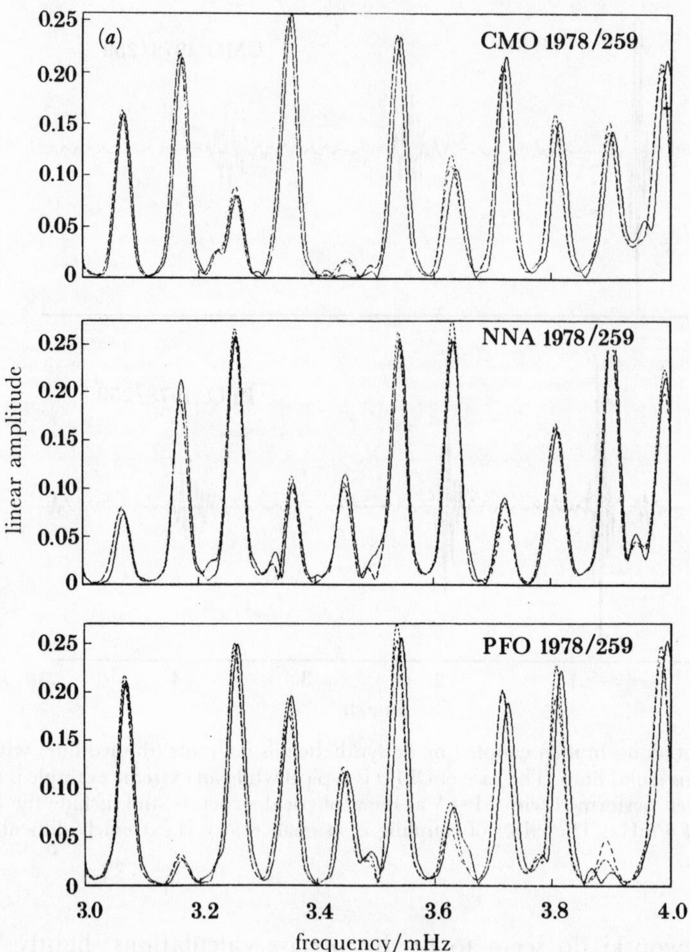

FIGURE 9a. For description see opposite.

to say, we do not do this well and it seems that M84A substantially underpredicts the magnitude of the variations in waveforms that we observe.

What is wrong with the model? There are several possibilities though experiments with synthetic seismograms suggest that the problem is probably not with elastic structure of low harmonic degree. In particular, enhancing the odd-order structure to increase the effect of along-branch coupling does not seem to help. Extreme lateral variations in the attenuation structure of the Earth may be a possibility or, more likely, the effect of higher order structure than that considered heretofore may be important. To investigate this latter possibility, we have expanded an example of a tectonic regionalization into spherical harmonics (Dziewonski & Steim 1982) and constructed a large suite of synthetic seismograms using various coupling schemes (this work was done in collaboration with Dr Ivan Henson). The regionalization, along with its spherical harmonic representation up to harmonic degree 40 is shown in figure 10. Because such models have sharp edges, the power in each harmonic degree falls off quite slowly (figure 11).

The relative r.m.s. residuals between the first ten hours of seismograms computed using different recipes are shown in table 2. Note that the numbers are roughly double those for

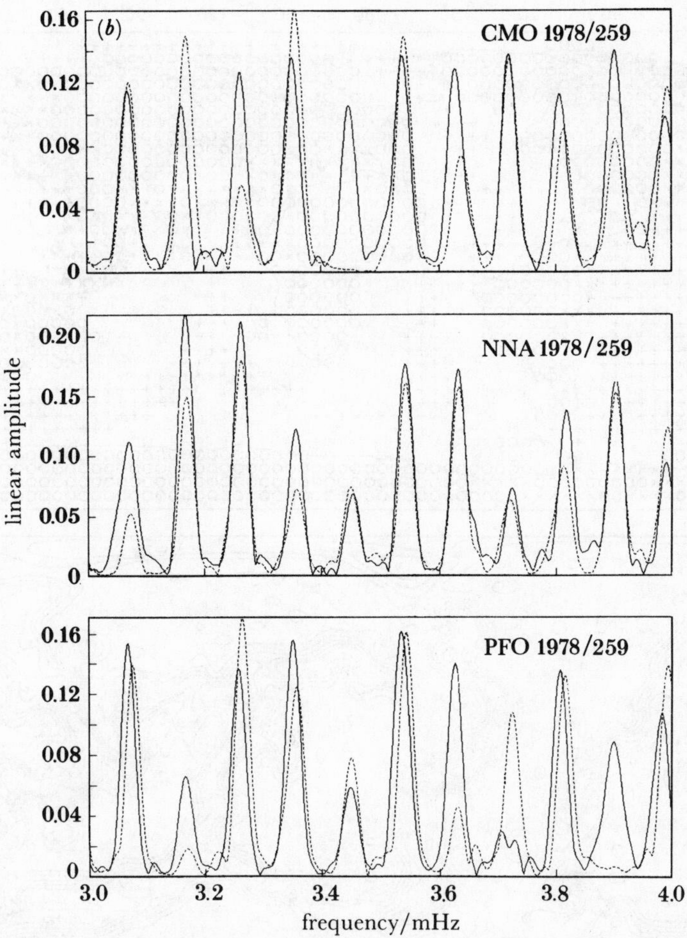

FIGURE 9. (*a*) Theoretical amplitude spectra of 20 hours of the recordings described in figure 8. The solid line is for a spherical Earth, the short dashed line assumes multiplets are isolated and the long dashed line includes along-branch coupling. The main effect of the aspherical model is to cause peak shifting and small amplitude perturbations. (*b*) As in (*a*), but showing the real data (solid line) in lieu of the spherical Earth seismogram. Note the strong amplitude anomalies that are not present in the theoretical calculation.

M84A and are more similar to the data. Also note that seismograms constructed with the method of Woodhouse & Dziewonski are sometimes worse fits to the most complete calculations than seismograms computed without any coupling at all. The higher order structure in this model has little effect on peak frequency shifts, but figure 12 does show that the apparent attenuation measurements are now quite strongly affected by the structure. Again, this is similar to the behaviour we observe in the data.

In summary, while a model like DS40 does not provide a quantitative fit to the data (indeed it doesn't even predict the degree 2 pattern seen in the peak frequency data), the presence of relatively low power, high harmonic degree structure does seem to produce qualitatively similar signals to those in the data. The presence of such higher-order structure may cause approximate theoretical representations to give biased answers particularly in the odd-order structure whose effects are only weakly apparent in the data (see also Park 1986).

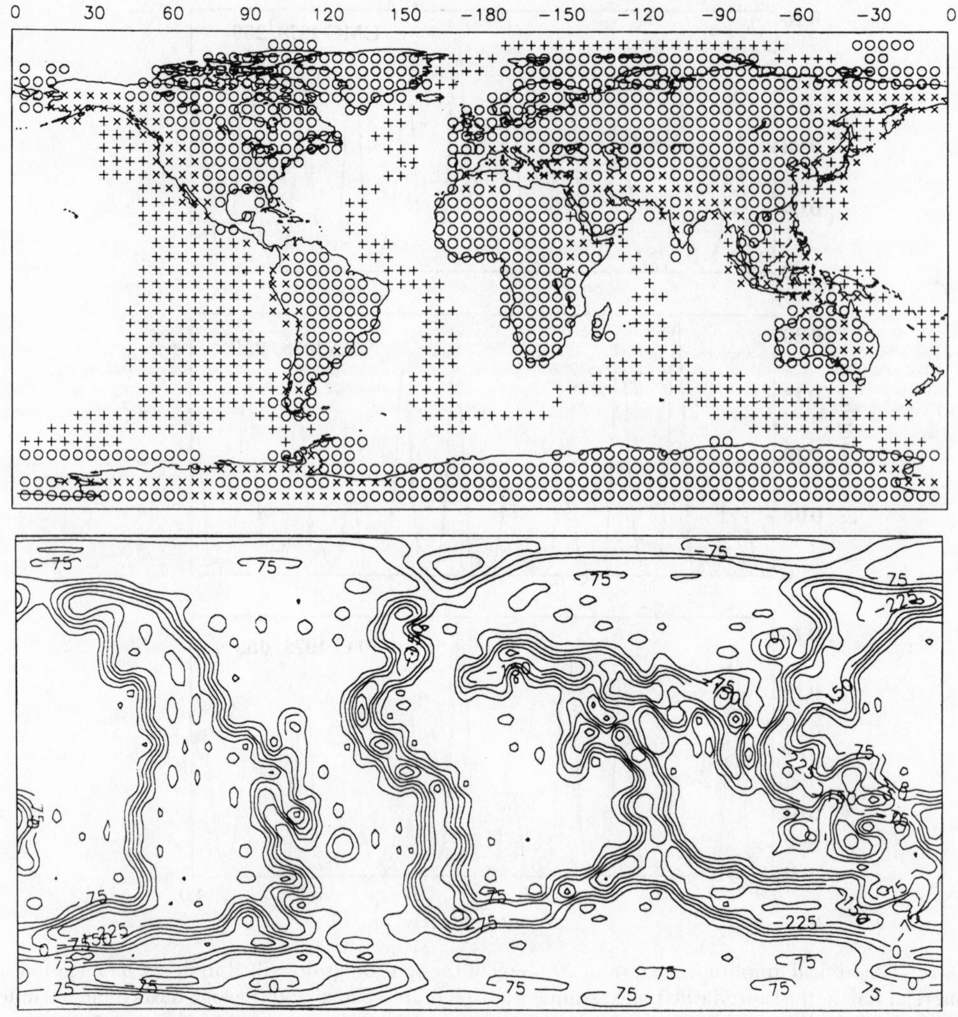

FIGURE 10. The tectonic regionalization of Dziewonski & Steim along with the expansion up to harmonic degree 40 of the lateral variation in shear velocity in one of the layers of this model.

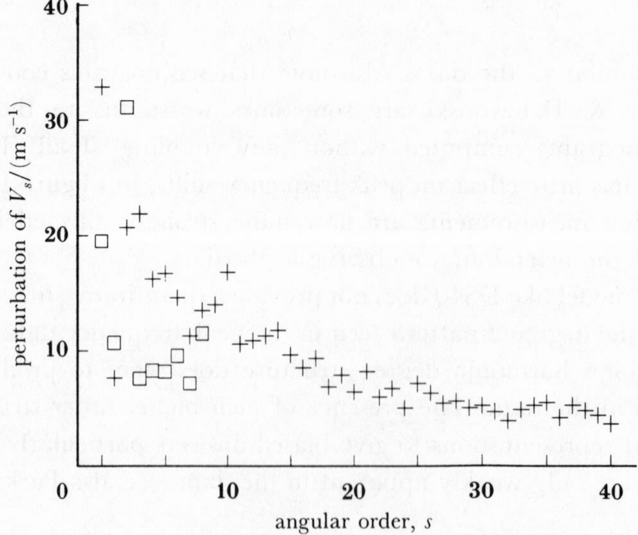

FIGURE 11. A comparison of the average amplitude in each harmonic degree of DS40 and M84A at a depth corresponding to the transition zone of the Earth. Note the dominant degree 2 part of M84A and the slow fall-off of the power in DS40.

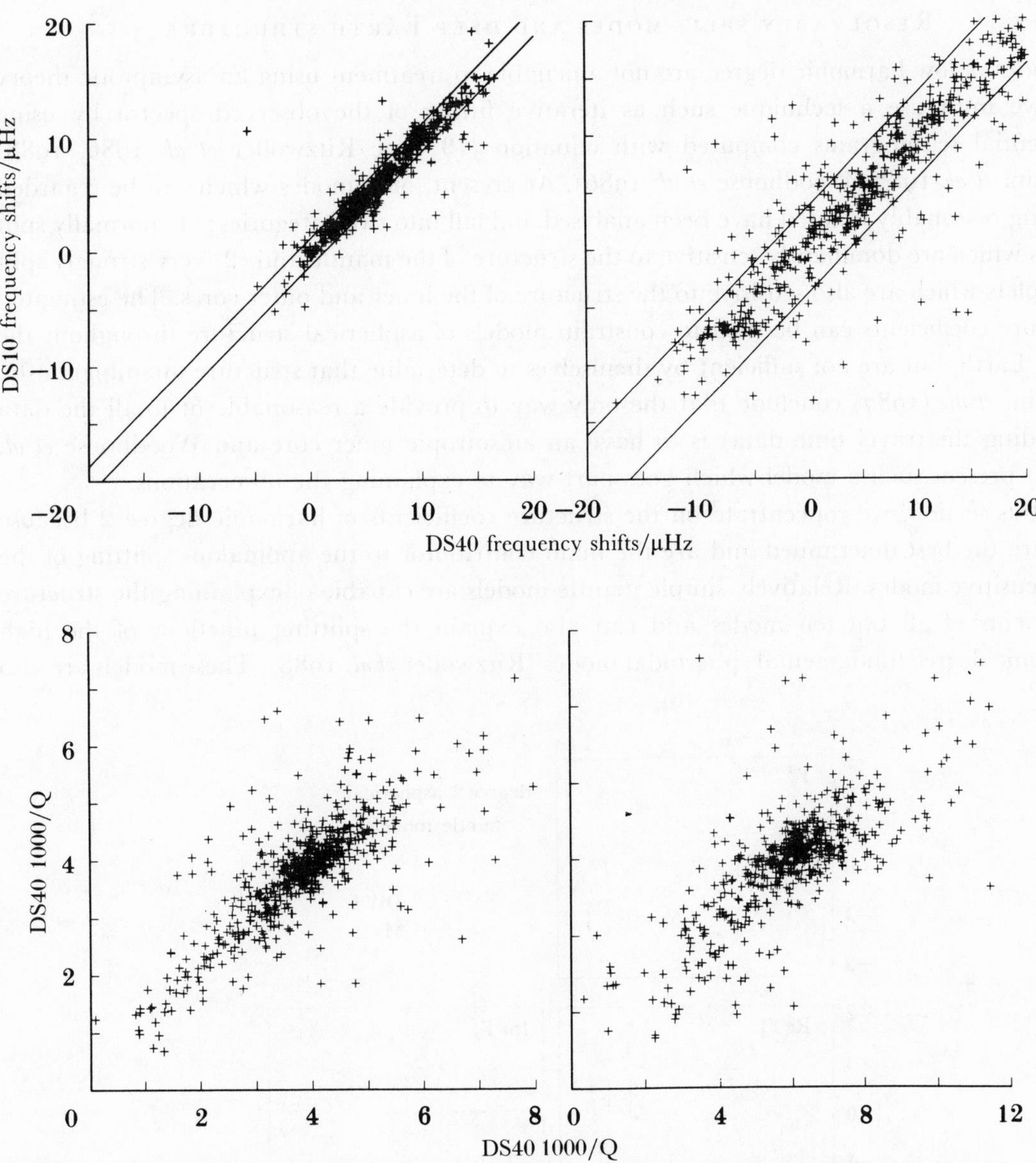

FIGURE 12. An illustration of the effect of higher-order structure on the centre frequencies and attenuation rates of ${}_0S_{23}$ (left column) and ${}_0S_{43}$ (right column). The scatter away from a straight line is solely due to structure of harmonic degree greater than 10. The centre frequencies show much less effect than the apparent attenuation implying that higher-order structure may cause some of our inability to model the attenuation data.

TABLE 2. AS FOR TABLE 1 BUT USING THE TECTONIC REGIONALIZATION

	coupled-5	coupled-3	WD84	isolated	peak shift	spherical
coupled-5	0	18.2	32.7	32.9	41.1	58.2
coupled-3	—	0	36.4	27.2	36.4	54.5
WD84	—	—	0	44.1	32.0	57.0
isolated	—	—	—	0	23.2	46.3
peak shift	—	—	—	—	0	43.4
spherical	—	—	—	—	—	0

Resolvably split modes and deep Earth structure

Modes of low harmonic degree are not amenable to treatment using an asymptotic theory and we must use a technique such as iterative fitting of the observed spectra by using differential seismograms computed with equation (19) (see Ritzwoller *et al.* 1986, 1988; Giardini *et al.* 1987; Woodhouse *et al.* 1986). At present, only modes which can be regarded as being reasonably isolated have been analysed and fall into two categories: (1) normally split modes which are dominantly sensitive to the structure of the mantle and (2) very strongly split multiplets which are also sensitive to the structure of the inner and outer cores. The estimated structure coefficients can be used to constrain models of aspherical structure throughout the whole Earth, but are not sufficient by themselves to determine that structure unambiguously. Giardini *et al.* (1987) conclude that the only way to provide a reasonable fit to all the data (including the travel time data) is to have an anisotropic inner core and Woodhouse *et al.* (1986) present such a model which goes part way to explaining the observations.

In this section, we concentrate on the structure coefficients of harmonic degree 2 because they are the best determined and are the main contributor to the anomalous splitting of the core-sensitive modes. Relatively simple mantle models are capable of explaining the structure coefficients of all but ten modes and can also explain the splitting functions of the high harmonic degree fundamental spheroidal modes (Ritzwoller *et al.* 1988). These models are also

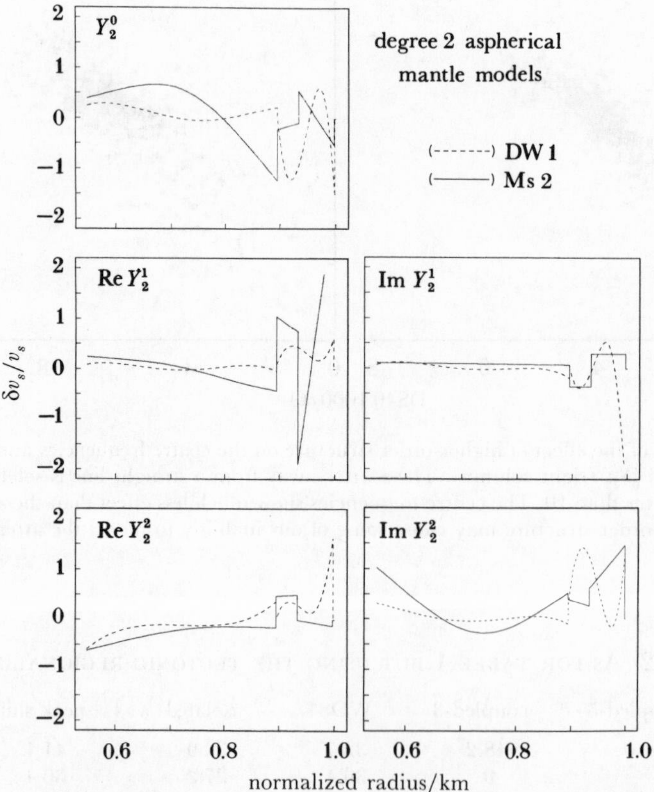

FIGURE 13. A degree 2 mantle model expressed as a relative perturbation in shear velocity derived from structure coefficients (solid line-Ms2) compared with a model which is a union of the upper mantle and lower mantle models of Dziewonski & Woodhouse (dashed line-DW1). Figure from Ritzwoller *et al.* (1988).

quite similar to the models derived from travel-time data (see, for example, Dziewonski 1984) except near the top of the lower mantle where the travel-time models have low sensitivity. The radial dependence of the harmonic degree 2 perturbations in shear velocity for one such model are shown in figure 13 and compared with a combination of models from the Harvard group. There are discrepancies in the upper mantle, but trade-offs with structure on internal discontinuities mean that it is difficult to construct a well-constrained model in this region. These models do not explain the structure coefficients of the ten anomalous core-sensitive modes. The most anomalous structure coefficient for those modes is c_2^0 which can be partially fit by introducing various kinds of anomalous structure into the core. It should be emphasized that the splitting of these modes is not a small signal as they have splitting widths between two and three times greater than that predicted by a rotating Earth in hydrostatic equilibrium. Figure 14 demonstrates that our mantle models do not fit these data though the addition of

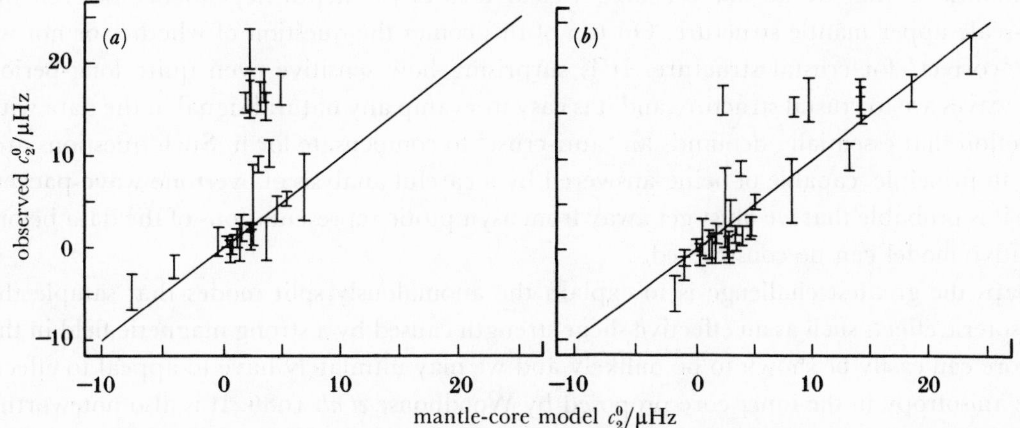

FIGURE 14. Observered c_2^0 coefficients plotted (*a*) against the predictions of the mantle model Ms2. (*b*) against the predictions of Ms2 which also includes some topography on the boundaries of the core and and isotropic heterogeneity in the inner core. Figure from Ritzwoller *et al.* (1988).

some topography on the boundaries of the outer core and a volumetric perturbation in the inner core go a long way to explaining the observations. The least well-fit modes are, unfortunately, the least well constrained by the observations so the poor fit to modes like ${}_3S_2$ may or may not be significant (Ritzwoller *et al.* 1988). While a model including anisotropy of the inner core can give a slightly better fit to some of the most anomalous modes (Woodhouse *et al.* 1986) it is likely that we will have to await an expansion of the data-set and an improvement in our knowledge of mantle structure before a definitive explanation for the splitting of these modes can be found.

CONCLUSIONS

The first round of modelling signals from aspherical structure in low-frequency data has produced some surprising results. In my opinion, the most surprising is that structure of harmonic degree 2 is dominant in the data and that the predictions of a rotating Earth in hydrostatic equilibrium are almost *never* fulfilled. This result is even more surprising when it is remembered that the surface of the Earth has a flattening that is very close to the theoretically predicted value yet even a 300 second Rayleigh wave thinks that the Earth is round rather than elliptical. We seem to be able to produce models of large-scale structure in the mantle that go some way to explaining the largest-scale signals but the construction of synthetic seismograms for these models show that we have a long way to go. In fact, it is probably true that our best models predict seismograms which are closer to the predictions of a purely spherical Earth model than to the real data. We have yet to tie down the reason for this inability to mode the actual waveforms in detail but the answer probably lies in a rather unglamorous combination of higher-order structure and complicated coupling effects.

As far as the dominant signals in the seismogram are concerned (the surface-wave equivalent modes) most investigators agree on structure coefficients of harmonic degree 2 and (to a lesser extent) structure of harmonic degree 6, though a model of aspherical structure predicting these effects is by no means uniquely determined from these data. In particular, there are severe trade-offs between volumetric perturbations and structure on internal discontinuities in the upper mantle so that we do not yet have a clear idea of the depth dependence of even the largest-scale upper mantle structure. On top of this comes the question of whether or not we should 'correct' for crustal structure. It is surprising how sensitive even quite long-period surface waves are to crustal structure and it is easy to swamp any natural signal in the data with a correction that essentially demands an 'anti-crust' to compensate for it. Such questions are, at least in principle, capable of being answered by a careful analysis of overtone wave-packets though it is probable that we must get away from asymptotic representations of the data before a definitive model can be constructed.

Perhaps the greatest challenge is to explain the anomalously split modes that sample the core. Esoteric effects such as an effective shear strength caused by a strong magnetic field in the outer core can easily be shown to be unlikely and we may ultimately have to appeal to effects like the anisotropy in the inner core proposed by Woodhouse *et al.* 1986. It is also noteworthy that recent attempts at the modelling of the spherically averaged structure by using very accurate degenerate mode frequencies have met with difficulties when it comes to finding adequate models of the deepest Earth structure and there may be something fundamentally wrong with our conception of what this region of the Earth looks like (F. Gilbert, personal communication). It is also true that our modelling efforts are based upon the analysis of relatively few modes and an expansion of the data-set is required before emphatic statements about the structure of the deep Earth can be made.

I thank Mike Ritzwoller, Ivan Henson, Mark Smith and Rudolf Widmer for their invaluable contributions to this work, of course, all errors and omissions remain my responsibility alone. This research was supported by National Science Foundation grants EAR-84-10369 and EAR-84-18471. Most of the computations were performed on the Cray X-MP at the San Diego Supercomputer Center.

References

Dahlen, F. A. 1979 *Geophys. Jl R. astr. Soc.* **58**, 1–33.
Dahlen, F. A. 1987 *Geophys. Jl R. astr. Soc.* **91**, 241–254.
Davis, J. P. 1987 *Geophys. Jl R. astr. Soc.* **88**, 693–722.
Dziewonski, A. M. 1984 *J. geophys. Res.* **89**, 5929–5952.
Dziewonski, A. M. & Steim, J. 1982 *Geophys. Jl R. astr. Soc.* **70**, 503–527.
Giardini, D., Li, X.-D. & Woodhouse, J. H. 1987 *Nature, Lond.* **325**, 405–409.

Gilbert, F. & Backus, G. 1966 *Geophysics* **31**, 326–332.
Gilbert, F. & Dziewonski, A. M. 1975 *Phil. Trans. R. Soc. Lond.* A **278**, 187–269.
Jordan, T. H. 1978 *Geophys. Jl R. astr. Soc.* **52**, 441–455.
Masters, G., Park, J. & Gilbert, F. 1983 *J. geophys. Res.* **88**, 10285–10298.
Masters, G. & Ritzwoller, M. 1988 In *Mathematical geophysics*, pp. 1–30. Dordrecht: D. Reidel.
Nakanishi, I. & Anderson, D. L. 1983 *J. geophys. Res.* **88**, 10267–10283.
Nakanishi, I. & Anderson, D. L. 1984 *Geophys. Jl R. astr. Soc.* **78**, 573–618.
Park, J. 1986 *J. geophys. Res.* **91**, 6441–6464.
Park, J. 1987 *Geophys. Jl R. astr. Soc.* **90**, 129–169.
Park, J. & Gilbert, F. 1986 *J. geophys. Res.* **91**, 7241–7260.
Ritzwoller, M., Masters, G. & Gilbert, F. 1986 *J. geophys. Res.* **91**, 10203–10228.
Ritzwoller, M., Masters, G. & Gilbert, F. 1988 *J. geophys. Res.* **93**, 6369–6396.
Romanowicz, B. 1987 *Geophys. Jl R. astr. Soc.* **90**, 75–100.
Schulten, K. & Gordon, R. 1975 *J. math Phys.* **16**, 1961–1970.
Smith, M. F. & Masters, G. 1989 *J. geophys. Res.* **94**, 1953–1976.
Tanimoto, T. 1987 *Geophys. Jl R. astr. Soc.* **89**, 713–740.
Woodhouse, J. H. & Dahlen, F. A. 1978 *Geophys. Jl R. astr. Soc.* **53**, 335–354.
Woodhouse, J. H. & Dziewonski, A. M. 1984 *J. geophys. Res.* **89**, 5953–5986.
Woodhouse, J. H. & Girnius, T. P. 1982 *Geophys. Jl R. astr. Soc.* **68**, 653–673.
Woodhouse, J. H. & Giardini, D. 1985 *Eos, Wash.* **66**, 300.
Woodhouse, J. H., Giardini, D. & Li, X.-D. 1986 *Geophys. Res. Lett.* **13**, 1549–1552.

Phil. Trans. R. Soc. Lond. A **328**, 351–363 (1989) 351
Printed in Great Britain

Fluctuations in the Earth's rotation and the topography of the core–mantle interface

By R. Hide, F.R.S.

Geophysical Fluid Dynamics Laboratory, Meteorological Office (Met 0 21), *Bracknell, Berkshire RG*12 2*SZ, U.K.*†

As arguments in favour of the notion that very slow convection in the highly viscous mantle is confined to the upper 700 km gradually weakened over the past 20 years, so geophysicists have increased their willingness to entertain the idea that significant horizontal variations in temperature and other structural parameters occur at all levels in the lower mantle. Concomitant density variations, including those caused by distortions in the shape of the core–mantle interface, would contribute substantially to long-wavelength features of the Earth's gravity field and also affect seismic travel times. The implied departures from axial symmetry in the thermal and mechanical boundary conditions thus imposed by deep mantle convection on the underlying low-viscosity liquid metallic core would affect not only *spatial* variations in the long-wavelength features of the main geomagnetic field (which is generated by dynamo action involving comparatively rapid chaotic magnetohydrodynamic flow in the core) but also *temporal* variations on all relevant timescales, from decades and centuries characteristic of the geomagnetic secular variation to tens of millions of years characteristic of changes in the frequency of polarity reversals.

Core motions should influence the rotation of the 'solid' Earth (mantle, crust and cryosphere), and in the absence of any quantitatively reasonable alternative line of attack, geophysicists have long supposed that irregular 'decade' fluctuations in the length of the day of about 5×10^{-3} s must be manifestations of angular momentum exchange between the core and mantle produced by time-varying torques at the core–mantle interface. The stresses responsible for these torques comprise (*a*) *tangential* stresses produced by viscous forces in the thin Ekman–Hartmann boundary layer just below the interface and also by Lorentz forces associated with the interaction of electric currents in the weakly conducting lower mantle with the magnetic field there, and (*b*) *normal* stresses produced largely by dynamical pressure forces acting on irregular interface topography (i.e. departures in shape from axial symmetry). The hypothesis that topographic stresses might provide the main contribution to the torque was introduced by the author in the 1960s and the present paper gives details of his recently proposed method for using Earth rotation and other geophysical data in a new test of the hypothesis. The method provides a scheme for investigating the consistency of the hypothesis with various combinations of 'models' of (*a*) motions in the outer reaches of the core based on geomagnetic secular variation data, and (*b*) core–mantle interface topography based on gravity and seismic data, thereby elucidating the validity of underlying assumptions about the dynamics and structure of the Earth's deep interior upon which the various 'models' are based. The scheme is now being applied in a complementary study carried out in collaboration with R. W. Clayton, B. H. Hager, M. A. Spieth and C. V. Voorhies.

1. Introduction

Recent advances in seismic tomography and the study of mantle convection are influencing progress in other areas of geophysics and geochemistry. Of particular interest in connection

† Present address: Robert Hooke Institute, Old Observatory, Clarendon Laboratory, Parks Road, Oxford OX1 3PU, U.K.

with the new international study of the Earth's deep interior (SEDI) now being organized under the auspices of the International Union of Geodesy and Geophysics will be improvements in our knowledge of the structure, composition and dynamics of the lower mantle and the liquid metallic core, where the geomagnetic field is produced by the self-exciting magnetohydrodynamic dynamo process. Historically, seismology has provided Earth scientists with crucial detailed knowledge about the radial variation of the Earth's density and elastic properties when averaged over spherical surfaces, but without having had much to say until fairly recently about asymmetric features of the structure of the Earth's deep interior (see Doornbos 1988; Silver & Carlson 1988; Woodhouse & Dziewonski, this Symposium). So those of us who in the 1960s and 1970s had been obliged to invoke such features to make sense of other types of geophysical data were unable then to call upon seismologists for detailed guidance. When, for instance, I was preparing a discussion (Hide 1970) of possible geophysical implications of observations of long-wavelength features of the Earth's main magnetic and gravitational fields and of the so-called 'decade' variations in the Earth's rate of rotation, my conclusions could not usefully be checked against seismic data. (An attempt on my part to use the inadequate seismic data then available to test the validity of hypothetical maps of core–mantle interface topography based on gravity data alone (Hide & Horai 1968) was amateurish, short-lived and thoroughly unsuccessful (but see Vogel 1960).)

I was viewing (so to speak) the Earth's interior through the eyes of a fluid dynamicist with some knowledge of the hydrodynamics and magnetohydrodynamics (MHD) of rapidly rotating fluids combined with a long-standing interest in the dynamo theory of the origin of the main geomagnetic field and its bearing on the interpretation of various properties of that field and its secular changes, including the striking discovery by palaeomagnetic workers that polarity reversals vary markedly in frequency over geological time (for references, see Jacobs 1975, 1984; Runcorn *et al.* 1982; Merrill & McElhinny 1983). In all realistic theoretical models, the specification of boundary conditions is crucial, and no thoroughly satisfactory model of the geodynamo can be formulated without due attention being paid to the thermal, mechanical and electromagnetic boundary conditions under which the full set of MHD equations would have to be solved. Much of course can be learned about dynamos through the study of highly simplified models (for references, see Moffatt 1978*a*; Jacobs 1987) but it was evident that essential features of a model capable of accounting in detail for the principal spatial and temporal characteristics of the main field as revealed by hard-won geomagnetic and palaeomagnetic data would be small but dynamically significant irregular departures from spherical symmetry in the boundary conditions.

To anyone who had thought seriously about these matters it seemed likely that irregular aspherical features of the shape of the core–mantle interface and the thermal field there could be produced and modulated over geological timescales by slow and time-varying convection in the highly viscous lower mantle. Having got off to a slow start, the idea is becoming accepted as providing a useful working hypothesis or paradigm, as several contributions to this very timely Discussion Meeting will attest and provide detailed references. It will be interesting to see what consensus emerges from all these contributions, and even more important will be our attempts to identify the crucial questions upon which attention must be focused in the immediate future. Recent progress made with the direct use of gravity, magnetic and seismic data will be discussed by other speakers (see papers by Hager & Richards, Gubbins, Woodhouse & Dziewonski, all in this Symposium). It is my allotted task to outline some of the implications for the Earth's deep interior of recent work on angular momentum transfer between the core and mantle, as evinced largely by the irregular variations in the length of the day that would be left when angular momentum exchange between the 'solid Earth' (mantle, crust and cryosphere) and overlying atmosphere and hydrosphere, moment of inertia changes due to internal processes, and various other effects had been allowed for (see §2 below).

A mechanical couple or torque acting between two bodies transfers angular momentum from one to the other at a rate proportional to the strength of the couple. The couple – a vector quantity with a precise definition – is caused by physical interactions of various kinds between the bodies, but there can obviously be other interactions that do not give rise to a mechanical couple. Now the imprecise term 'core–mantle coupling', usually qualified by the adjective 'viscous', 'electromagnetic' or 'topographic' (see equation (2.1)), is widely used in the extensive literature on the interpretation of the decade variations in the length of the day as being manifestations of angular momentum exchange between the core and mantle (but the more precise term 'torquing' is seen only occasionally). This usage turns out to be unfortunate, for it may have led to the misunderstandings evident in the writings of authors who, by interpreting the term 'coupling' as embracing a wider class of interactions than just those that directly produce mechanical 'couples', have promoted a false dichotomy concerning the relative importance of thermal and topographic contributions to interactions between the core and mantle. Deep convection in the highly viscous mantle would produce substantial horizontal temperature variations near the core–mantle interface, and concomitant stresses in the mantle would deform the core–mantle interface, thereby producing irregular aspherical topography of the interface. Through their effects on motions in the fluid core, these fields of temperature and topography at the bottom of the mantle would affect the behaviour of the main geomagnetic field on all relevant timescales, from those characteristic of the geomagnetic secular variation (decades and centuries) to those associated with mantle convection and geological processes (millions of years and longer).

Detailed theoretical studies of these interactions will be of great interest in geophysics, just as investigations of both mean *and* time-varying distortions of atmospheric motions due to the presence of mountains and imposed irregular horizontal temperature variations at the Earth's surface are important in dynamical meteorology (see e.g. Kalnay & Mo 1986). Unlike the field of topography at the core–mantle interface, however, the associated temperature field in the lower mantle cannot contribute directly to the mechanical couple acting at the interface, although together with the ambient magnetic field, they are likely to produce important indirect effects. In my own work on decade variations in the length of the day I introduced the idea and emphasized the likely importance of topographic contributions to that couple (Hide 1969). But I have never dismissed thermal interactions between the mantle and core as being geophysically unimportant; on the contrary, the first discussion of the likely significance of such interactions was possibly given in a paper of mine (Hide 1967; see also Hide & Malin 1971; Malin & Hide 1982; Bloxham & Gubbins 1987). So I take this opportunity to reiterate my long-held view that the investigation of the dynamical processes involved in these various interactions and of their geophysical consequences, as in the important work reviewed by Gubbins (this Symposium), deserves high priority in theoretical work on the structure and dynamics of the Earth's deep interior.

2. Torques at the core–mantle interface

Irregular fluctuations in the rate of rotation of the solid Earth and also in the alignment of the rotation axis relative to the figure axis are produced by two principal agencies. These are: (*a*) changes in the inertia tensor of the solid Earth associated with the melting of ice, mantle convection, earthquakes, deformations associated with a variety of applied stresses, etc.; and (*b*) angular momentum transfer between the solid Earth and the fluid regions (liquid core, hydrosphere and atmosphere) with which the solid Earth is in contact. Of direct relevance in the investigation of the structure and dynamics of the liquid metallic core and lower mantle are the so-called 'decade' fluctuations in the length of the day (LOD) of up to about 5×10^{-3} s on timescales upwards of a few years and associated polar motion. The atmosphere is now known to be responsible for observed LOD fluctuations of about 1×10^{-3} s on timescales ranging from a few days to a few years and also for much if not all the observed polar motion on these timescales. But in the face of quantitative difficulties involved in accounting for the comparatively large but slower 'decade' fluctuations in terms of atmospheric (or oceanic) processes or of plausible changes in the inertia tensor of the solid Earth on relevant timescales, geophysicists have argued that the fluctuations are best regarded as manifestations of angular momentum transfer between the core and mantle, caused by dynamical stresses exerted on the mantle by irregular ('chaotic') flow in the core with typical speeds of about 3×10^{-4} m s^{-1} (for extensive references, see Munk & Macdonald 1960; Hide 1977, 1984, 1985; Lambeck 1980; Rochester 1984; Melchior 1986; Cazenave 1986; Dickey *et al.* 1986; Chao & Gross 1987; Hinderer *et al.* 1987; Moritz & Mueller 1987; Jault *et al.* 1988; Merriam 1988; Vondrák & Pejović 1988; Wahr 1988).

When integrated over the whole of the core–mantle boundary (CMB), these stresses give rise to a net couple

$$\boldsymbol{L}^*(t) \equiv \iint_{\mathrm{CMB}} \boldsymbol{r} \times [\boldsymbol{F}_{\mathrm{V}} + \boldsymbol{F}_{\mathrm{E}} + \boldsymbol{F}_{\mathrm{T}}]\,\mathrm{d}A. \tag{2.1}$$

Here t denotes time, $\boldsymbol{r}$ is the position vector of a general point referred to the Earth's centre of mass, $\mathrm{d}A$ the area element of the CMB, $\boldsymbol{F}_{\mathrm{V}}(t)$ the stress associated with shearing motions in the viscous boundary layer, $\boldsymbol{F}_{\mathrm{E}}(t)$ the electromagnetic stress associated with the Lorentz force $\boldsymbol{j} \times \boldsymbol{B}$ (where $\boldsymbol{j}$ is the electric current density and $\boldsymbol{B}$ the magnetic field) and $\boldsymbol{F}_{\mathrm{T}}(t)$ the 'topographic stress' due to the action of normal pressure forces on bumps (including the equatorial bulge) in the shape of the core–mantle interface. The 'axial' component of $\boldsymbol{L}^*$ produces changes in the length of the day; the 'equatorial components' move the pole of rotation of the solid Earth relative to its axis of figure (see equation (2.9) below).

The viscous contribution $\boldsymbol{F}_{\mathrm{V}}$ would be negligible on all but the most extreme assumptions about viscous forces in the core (see Bullard *et al.* 1950). Much more promising but still controversial owing to both qualitative and quantitative difficulties is the electromagnetic contribution, which has been the subject of a considerable number of theoretical studies since it was first considered by Bullard *et al.* (1950) (for references, see Rochester 1984; Paulus & Stix 1986; Roberts 1989). The theory of topographic coupling is less straightforward (Hide 1969, 1977; Anufriev & Braginsky 1977; Eltayeb & Hassan 1979; Moffatt 1978*b*; Roberts 1989), but rough dynamical arguments show that bumps no more than 10^3 m in vertical amplitude might suffice to make the axial component of $\boldsymbol{L}^*(t)$ large enough to account for the magnitude of the observed decade LOD changes (Hide 1969).

The determination of the topographic contribution $\boldsymbol{L}(t)$ (say) to $\boldsymbol{L}^*(t)$ (see equation (2.1)) from geophysical data using a method proposed by Hide (1986) is currently the subject of a complementary study (see §4 below). $\boldsymbol{L}(t)$ is given by the equation

$$\boldsymbol{L}(t) \equiv \iint_{\text{CMB}} \boldsymbol{r} \times \boldsymbol{F}_{\text{T}} \,\mathrm{d}A = \iint_{\text{CMB}} \boldsymbol{r} \times p_{\text{s}} \boldsymbol{n} \,\mathrm{d}A, \tag{2.2}$$

where $\boldsymbol{n}$ is the outwardly directed unit vector normal to the core–mantle boundary and p_{s} is the dynamic pressure associated with core motions just below the CMB. If the CMB is the locus of points where

$$r = r(\theta, \phi) = c + h(\theta, \phi), \tag{2.3}$$

(where c is the mean radius of the core (3480 km), θ co-latitude and ϕ longitude) then

$$\boldsymbol{L} = -c^2 \int_0^{2\pi} \int_0^{\pi} (\boldsymbol{r} \times p_{\text{s}} \nabla_{\text{s}} h) \sin\theta \,\mathrm{d}\theta \,\mathrm{d}\phi, \tag{2.4}$$

where $\nabla_{\text{s}} \equiv c^{-1}(\hat{\boldsymbol{\theta}}\, \partial/\partial\theta + \hat{\boldsymbol{\phi}} \operatorname{cosec}\theta\, \partial/\partial\phi)$ if $\hat{\boldsymbol{\theta}}$ and $\hat{\boldsymbol{\phi}}$ are unit vectors in the directions of increasing θ and ϕ respectively. Now it is readily shown that the surface integral of $\boldsymbol{r} \times \nabla_{\text{s}}(hp_{\text{s}})$ vanishes, so that equation (2.4) can be expressed in the equivalent and more convenient form

$$\boldsymbol{L}(t) = c^2 \int_0^{2\pi} \int_0^{\pi} (\boldsymbol{r} \times h \nabla_{\text{s}} p_{\text{s}}) \sin\theta \,\mathrm{d}\theta \,\mathrm{d}\phi. \tag{2.5}$$

An expression for $h\nabla_{\text{s}} p_{\text{s}}$ is obtainable by considering the hydrodynamical equation of motion, which throughout most of the core can be approximated by

$$2\rho \bar{\boldsymbol{\Omega}} \times \boldsymbol{u} + \nabla p - \boldsymbol{g}\rho \approx \boldsymbol{j} \times \boldsymbol{B}, \tag{2.6}$$

where ρ denotes density, p the total pressure, $\boldsymbol{u}$ the eulerian flow velocity in a frame of reference that rotates with angular velocity $\bar{\boldsymbol{\Omega}}$ relative to an inertial frame, and $\boldsymbol{g}$ is the acceleration due to gravity and centrifugal effects (see Hide 1986, equation (2.1)). It has long been recognized that over length scales comparable with the radius of the core, Coriolis effects on fluid motions there, as represented by the term $2\rho\boldsymbol{\Omega} \times \boldsymbol{u}$ in equation (2.6), are many orders of magnitude greater than the neglected relative acceleration terms $\rho\, \partial \boldsymbol{u}/\partial t$ and $\rho(\boldsymbol{u} \cdot \nabla)\, \boldsymbol{u}$ and the viscous term. The magnitude of the Lorentz term $\boldsymbol{j} \times \boldsymbol{B}$ on the right-hand side of equation (2.6) is uncertain, for we do not know the strength of the magnetic field within the core of the Earth. It has been argued that $\boldsymbol{B}$ is unlikely on average to exceed that value for which 'magnetostrophic' balance obtains, when $2\rho\bar{\boldsymbol{\Omega}} \times \boldsymbol{u}$ and $\boldsymbol{j} \times \boldsymbol{B}$ have the same order of magnitude (see Hide & Roberts 1978). Moreover, it is a fortunate circumstance that owing to the relatively poor electrical conductivity of the mantle the magnitudes of $\boldsymbol{j}$ and $\boldsymbol{B}$ near the core–mantle interface will be much less, by about a factor of 10, than the magnetostrophic value, the corresponding magnitude of the right-hand side of equation (2.6) being about 10^{-2} times that of $2\rho\bar{\boldsymbol{\Omega}} \times \boldsymbol{u}$. This implies that motions in the outer reaches of the core (but *not* within the viscous boundary layer of no more than about 1 m in thickness just below the CMB) may be characterized by geostrophic balance between Coriolis acceleration and the dynamic pressure gradient (see Le Mouël 1984), that is to say

$$2\rho\bar{\boldsymbol{\Omega}} \times \boldsymbol{u} \approx -\nabla p + \boldsymbol{g}\rho, \tag{2.7}$$

the curl of which gives $(2\bar{\boldsymbol{\Omega}}\cdot\nabla)\,\boldsymbol{u} \approx \rho^{-2}\nabla\rho\times\nabla p \approx \rho^{-1}\boldsymbol{g}\times\nabla\rho$. Taking $|2\bar{\boldsymbol{\Omega}}| \approx 10^{-4}\ \mathrm{s^{-1}}$, $\rho \approx 10^{4}\ \mathrm{kg\ m^{-3}}$ and $|\boldsymbol{u}| \approx 3\times10^{-4}\ \mathrm{m\ s^{-1}}$ gives $|\nabla p - \boldsymbol{g}\rho| \approx 3\times10^{-4}\ \mathrm{N\ m^{-3}}$, corresponding to pressure variations of about 300 N $\mathrm{m^{-2}}$ (3 mbar) over horizontal distances of the order of 10^{6} m and to concomitant seismologically undetectable but dynamically highly significant fractional density and pressure variations of order $|2\bar{\boldsymbol{\Omega}}\times\boldsymbol{u}|/|\boldsymbol{g}| = 3\times10^{-9}$!

Equations (2.7) and (2.5) provide the basis of the above-mentioned method for evaluating the torque on the mantle due to topographic stresses at the CMB. Denote by $\boldsymbol{u}_{\mathrm{s}}$ the eulerian flow velocity in the free stream just below the viscous boundary layer at the CMB, and by $(u_{\mathrm{s}}, v_{\mathrm{s}}, w_{\mathrm{s}})$ the (r,θ,ϕ) components of $\boldsymbol{u}_{\mathrm{s}}$, where u_{s} is typically so much smaller in magnitude than v_{s} and w_{s} that it can safely be set equal to zero. If $\nabla_{\mathrm{s}} p_{\mathrm{s}}$ is the corresponding value of the horizontal pressure gradient and $\bar{\rho}_{\mathrm{s}}$ the horizontally averaged value of $\rho(r\approx c)$ then by equation (2.7) we have $(v_{\mathrm{s}}, w_{\mathrm{s}}) = (2\bar{\rho}\bar{\Omega}c\cos\theta)^{-1}(-\operatorname{cosec}\theta\,\partial p_{\mathrm{s}}/\partial\phi, \partial p_{\mathrm{s}}/\partial\theta)$, so that v_{s} and $\partial p_{\mathrm{s}}/\partial\theta$ vanish on the Equator.

By equations (2.5) and (2.7),

$$\boldsymbol{L}(t) = 2\bar{\rho}_{\mathrm{s}}\,\Omega c^{3}\int_{0}^{2\pi}\int_{0}^{\pi} h(\theta,\phi)\,\boldsymbol{u}_{\mathrm{s}}(\theta,\phi,t)\sin\theta\cos\theta\,\mathrm{d}\theta\,\mathrm{d}\phi \tag{2.8}$$

where on the timescales of interest here, $\boldsymbol{u}_{\mathrm{s}}$ depends on t but h does not. Consider a set of body-fixed axes x_i $(i = 1,2,3)$ aligned with the principal axes of the solid Earth and rotating with angular velocity $\omega_i(t)$ about its centre of mass. By equation (2.8), the topographic torque $\boldsymbol{L}(t)$ has three components $L_i(t)$ $(i = 1,2,3)$ in this system given by

$$L_i(t) = 2\bar{\rho}_{\mathrm{s}}\bar{\Omega}c^{3}\int_{0}^{2\pi}\int_{0}^{\pi} h\{v_{\mathrm{s}}\cos\theta\cos\phi - w_{\mathrm{s}}\sin\phi,\ v_{\mathrm{s}}\cos\theta\sin\phi + w_{\mathrm{s}}\cos\phi,\ -v_{\mathrm{s}}\sin\theta\}\sin\theta\cos\theta\,\mathrm{d}\theta\,\mathrm{d}\phi. \tag{2.9}$$

The axial component L_3 changes the length of the day $(2\pi/\omega_3)$ and the equatorial components L_1 and L_2 move the pole of rotation, whose position relative to the figure axis is specified by ω_1 and ω_2 (see equations (3.36)–(3.39) below).

The dynamical processes that produce within the fluid regions of the Earth the stresses responsible for transferring angular momentum to the solid Earth are by no means fully understood. Their elucidation will require much further study by theoreticians and experimentalists concerned with the hydrodynamics and magnetohydrodynamics of rotating fluids. Discussion of the fascinating problems involved lies beyond the scope of the present paper (see Hide 1969, 1977; Anufriev & Braginsky 1977; Moffatt 1978*b*; Eltayeb & Hassan 1979; Roberts 1989), but a general observation on the strategy to be followed in this line of research is worth making. It is important not to oversimplify models of the interaction of core motions with the core–mantle interface, and in this connection, the case when effects due to nonlinear advection of momentum, buoyancy forces, magnetic fields and viscosity are all neglected is instructive. Then, in accordance with well-known results concerning the 'spin-up' of fluids in irregular containers (see, for example, Greenspan 1968), the dynamical pressure field would be symmetric about the equatorial plane and vary with θ and ϕ in such a way that the L_3 vanishes! This is understandable, for when angular momentum is exchanged between the solid Earth and its fluid regions, the total rotational kinetic energy of the whole system must change if the speed of rotation of the solid Earth changes, even if the total angular momentum of the whole system remains constant. So successful models of the interaction processes must include mechanisms capable of producing transformations between rotational kinetic energy and other forms of energy (non-rotational kinetic energy, gravitational potential energy, magnetic energy and thermal energy) through the action of some or all the above-mentioned agencies.

3. Dynamics of the Earth's rotation

In this brief account of non-rigid body rotation, the equations needed for the study of the variable rotation of the Earth are derived in a form appropriate to this discussion of core–mantle interactions. Euler's dynamical equations describing the response of the 'whole Earth' (namely the 'solid Earth' plus the underlying inner (solid) core and outer (liquid) core and the overlying hydrosphere and lower and upper atmosphere) to an externally applied torque $\hat{L}_i$ are the following:

$$\mathrm{d}H_i/\mathrm{d}t + \epsilon_{ijk}\,\omega_j\,H_k = \hat{L}_i, \tag{3.1}$$

the usual convention being used for repeated suffices. Here ϵ_{ijk} is the alternating tensor (with values 0 or ± 1), $\mathrm{d}/\mathrm{d}t$ is the time-derivative in the rotating frame, and H_i, the absolute angular momentum, is given by

$$H_i(t) \equiv I_{ij}(t)\,\omega_j(t) + h_i(t). \tag{3.2}$$

The quantity I_{ij} is the variable inertia tensor defined by the volume integral (taken over the whole Earth)

$$I_{ij} \equiv \oiiint \rho\,(x_k\,x_k\,\delta_{ij} - x_i\,x_j)\,\mathrm{d}V, \tag{3.3}$$

where $\mathrm{d}V$ denotes volume element, δ_{ij} is the Kronecker delta (with values 0 and 1) and

$$h_i \equiv \oiiint \rho\epsilon_{ijk}\,x_j\,u_k\,\mathrm{d}V \tag{3.4}$$

is the angular momentum due to motion u_i relative to the axes x_i. Substitution in (3.1) gives the Liouville equation

$$\mathrm{d}/\mathrm{d}t(I_{ij}\,\omega_j + h_i) + \epsilon_{ijk}\,\omega_j(I_{kl}\,\omega_l + h_k) = \hat{L}_i, \tag{3.5}$$

further details of which, including expressions for all three components, can be found in various texts (see, for example, Munk & MacDonald 1960; Lambeck 1980).

Because the rotation of the Earth departs only slightly from steady rotation about the polar axis of figure, we write

$$\omega_i = (\omega_1, \omega_2, \omega_3) = \Omega(m_1, m_2, 1 + m_3) \tag{3.6}$$

where Ω is the mean rotational speed of the Earth, $0.729\,211\,5 \times 10^{-4}$ radians per sidereal second. The quantities m_1, m_2 and m_3 are very much less than unity and $|\dot{m}_i| \equiv |\mathrm{d}m_i/\mathrm{d}t| \ll \Omega$. The motion of the pole is given by $(m_1(t), m_2(t))$ and it will be convenient to define the quantity

$$\boldsymbol{m} \equiv m_1 + \mathrm{i}m_2 \tag{3.7}$$

where $\mathrm{i} \equiv \sqrt{-1}$. If $\Delta\Lambda(t) \equiv \Lambda(t) - \Lambda_0$, the difference between the instantaneous length of the day $2\pi/\omega_3$ and its average value $2\pi/\Omega$, then by equation (3.6),

$$m_3(t) = -\Delta\Lambda(t)/\Lambda_0. \tag{3.8}$$

In the same spirit, we now treat the quantities I_{ij} and h_i. Define

$$I_{ij} \equiv I_{ij}^{(c)} + I_{ij}^{(m)} + I_{ij}^{(a)}, \tag{3.9}$$

$$h_i \equiv h_i^{(c)} + h_i^{(m)} + h_i^{(a)}, \tag{3.10}$$

$$A \equiv A^{(c)} + A^{(m)} + A^{(a)}, \tag{3.11}$$

and
$$C \equiv C^{(c)} + C^{(m)} + C^{(a)}, \tag{3.12}$$

where the superscripts (c), (m) and (a) refer respectively to the whole core (inner and outer), the 'solid Earth' (mantle, crust and cryosphere), and the outer fluid layers (comprising the hydrosphere and the upper and lower atmosphere). $A^{(c)}, C^{(c)}$, etc., as defined by equations (3.11) and (3.12) are the principal moments of inertia of the regions to which they refer. It is readily shown by considering the orders of magnitude of the various terms in these equations that

$$C - A \approx 3 \times 10^{-3}\, C \ll C; \quad A^{(a)} \approx C^{(a)} \approx 3 \times 10^{-4}\, A \ll A;$$

$$A^{(c)} \approx C^{(c)} \approx 0.1\, C^{(m)} \ll C^{(m)}; \quad |h_i^{(c)}| \gtrsim |h_i^{(a)}| \gg h_i^{(m)}$$

and
$$|h_i^{(c)} + h_i^{(a)}| \approx 5 \times 10^{-8}\, \Omega C \ll \Omega C. \tag{3.13}$$

So we can write
$$h_i = h_i^{(c)} + h_i^{(a)}, \tag{3.14}$$

$$A = A^{(c)} + A^{(m)}, \quad C = C^{(c)} + C^{(m)} \tag{3.15}$$

and
$$I_{ij} = \begin{pmatrix} A & 0 & 0 \\ 0 & A & 0 \\ 0 & 0 & C \end{pmatrix} + \Delta I_{ij}, \tag{3.16}$$

with corresponding separate expressions for $I_{ij}^{(c)}$ and $I_{ij}^{(m)}$, if

$$\Delta I_{ij} \equiv \Delta I_{ij}^{(c)} + \Delta I_{ij}^{(m)} + I_{ij}^{(a)}. \tag{3.17}$$

Adopting a perturbation approach by combining equations (3.6) and (3.16) with equation (3.5) and neglecting second-order quantities, we find

$$\sigma_r^{-1}\, \dot{m}_1 + m_2 = \psi_2, \tag{3.18}$$

$$\sigma_r^{-1}\, \dot{m}_2 - m_1 = -\psi_1, \tag{3.19}$$

$$\dot{m}_3 = \dot{\psi}_3, \tag{3.20}$$

where
$$\sigma_r \equiv \Omega(C-A)/A \tag{3.21}$$

is the 'rigid body' frequency corresponding to a period of about 10 months. The non-homogeneous forcing terms are known as the excitation functions given by

$$\psi_1 \equiv [\Omega^2 \Delta I_{13} + \Omega \Delta \dot{I}_{23} + \Omega h_1 + \dot{h}_2 - \hat{L}_2]/\Omega^2(C-A), \tag{3.22}$$

$$\psi_2 \equiv [\Omega^2 \Delta I_{23} - \Omega \Delta \dot{I}_{13} + \Omega h_2 - \dot{h}_1 + \hat{L}_1]/\Omega^2(C-A), \tag{3.23}$$

and
$$\psi_3 \equiv \left[\frac{-\Omega \Delta I_{33} - h_3}{\Omega C}\right] + \frac{1}{\Omega C}\int_0^t \hat{L}_3(\tau)\, \mathrm{d}\tau, \tag{3.24}$$

where τ is a dummy variable.

The solution of equations (3.18) and (3.19) for the equatorial components m_1 and m_2 is

$$\boldsymbol{m}(t) = \exp(\mathrm{i}\sigma_r t)\left\{\boldsymbol{m}(0) - \mathrm{i}\sigma_r \int_0^t \boldsymbol{\psi}(\tau)\exp(-\mathrm{i}\sigma_r \tau)\,\mathrm{d}\tau\right\} \tag{3.25}$$

(see equation (3.7)), where $\boldsymbol{\psi} \equiv \psi_1 + \mathrm{i}\psi_2$ satisfies

$$\boldsymbol{\psi} = [\Omega^2 \Delta \boldsymbol{I} - \mathrm{i}\Omega\Delta\dot{\boldsymbol{I}} + \Omega\boldsymbol{h} - \mathrm{i}\dot{\boldsymbol{h}} + \mathrm{i}\hat{\boldsymbol{L}}]/\Omega^2(C-A) \tag{3.26}$$

if $$\Delta\boldsymbol{I} \equiv \Delta I_{13} + \mathrm{i}\Delta I_{23}, \quad \boldsymbol{h} \equiv h_1 + \mathrm{i}h_2$$

and $$\hat{\boldsymbol{L}} \equiv \hat{L}_1 + \mathrm{i}\hat{L}_2. \tag{3.27}$$

The corresponding solution of equation (3.20) for the axial component is

$$m_3(t) = \psi_3(t) + \text{constant}. \tag{3.28}$$

In the absence of external torques (i.e. when $\hat{L}_i = 0$) equations (3.23) and (3.27) with appropriate values of $\boldsymbol{\psi}$ and ψ_3 give $\boldsymbol{m}$ and m_3 for the case when the total angular momentum of the whole Earth is conserved. Conservation of total angular momentum about the x_3 axis, as expressed by equation (3.28) with $\hat{L}_3 = 0$, gives

$$\Omega[C^{(c)} + C^{(m)}](1+m_3) + \Omega[\Delta I_{33}^{(c)} + \Delta I_{33}^{(m)} + \Delta I_{33}^{(a)}] + h_3^{(c)} + h_3^{(a)} = \text{constant}, \tag{3.29}$$

with corresponding but rather more complicated expressions for $\boldsymbol{m}$. Write

$$\boldsymbol{\psi} \equiv \boldsymbol{\psi}^{(c)} + \boldsymbol{\psi}^{(m)} + \boldsymbol{\psi}^{(a)} \quad \text{and} \quad \psi_3 \equiv \psi_3^{(c)} + \psi_3^{(m)} + \psi_3^{(a)}, \tag{3.30}$$

where $\boldsymbol{\psi}^{(c)}$ and $\psi_3^{(c)}$ comprise all the terms in $\boldsymbol{\psi}$ and ψ_3 respectively involving $\Delta I_{ij}^{(c)}, h_i^{(c)}$ etc., and likewise for $\boldsymbol{\psi}^{(m)}$, $\psi_3^{(m)}$, $\boldsymbol{\psi}^{(a)}$ and $\psi_3^{(a)}$. Equations (3.25) and (3.28) can then be written as follows

$$\boldsymbol{m}(t) = \exp(\mathrm{i}\sigma_r t)\left\{\boldsymbol{m}(0) - \mathrm{i}\sigma_r \int_0^t [\boldsymbol{\psi}^{(c)}(\tau) + \boldsymbol{\psi}^{(m)}(\tau) + \boldsymbol{\psi}^{(a)}(\tau)]\exp(-\mathrm{i}\sigma_r \tau)\,\mathrm{d}\tau\right\} \tag{3.31}$$

and $$m_3(t) = \psi_3^{(c)}(t) + \psi_3^{(m)}(t) + \psi_3^{(a)}(t) + \text{constant} \tag{3.32}$$

respectively. As the angular momentum H_i of the whole Earth remains constant when $\hat{L}_i = 0$, changes in the angular velocity $\omega_i = \Omega(m_1, m_2, 1+m_3)$ of the solid Earth can only be produced by changes in the inertia tensor $I_{ij}^{(m)}$ and the exchange of angular momentum with the underlying core and the overlying hydrosphere and atmosphere.

So far as the evaluation of angular momentum exchange is concerned, the method to be adopted for determining the excitation functions $\psi_i^{(a)}$ and $\psi_i^{(c)}$ (and the contributions to $\psi_i^{(m)}$ that they produce because the solid Earth is not perfectly rigid) depends on the data available. In the so-called 'torque' approach, the rate of change of angular momentum $I_{ij}^{(m)}\,\omega_i$ of the solid Earth is directly related to the torques exerted upon its boundaries as a consequence of the motions within the fluid regions with which it is in contact. In the alternative 'angular momentum' approach, $\mathrm{d}(I_{ij}^{(m)}\,\omega_j)/\mathrm{d}t$ is taken to be equal and opposite (when $\hat{L}_i = 0$) to the rate of change of the total angular momentum of the fluid regions. Now the dominant contribution to $\psi_i^{(a)}$ comes from the tropospheric and lower stratospheric regions of the atmosphere, for which fairly abundant accurate wind and pressure observations are available, but surface stresses are not easy to evaluate. So the angular momentum rather than the torque approach is appropriate in the investigation of atmospheric excitation of changes in the Earth's rotation, and it is being used with great success in the study of short-term variations of all three components of ω_i of atmospheric origin and the determination by subtraction of non-meteorological contributions to $\dot{\omega}_i$ (for references see Hide 1984, 1985; Cazenave 1986; Dickey *et al.* 1986; Wahr 1988). However, for changes in ω_i brought about by the interaction between the core and the solid Earth, the angular momentum approach is not practicable because we have insufficient information about motions in the main body of the core (but see Jault *et al.* 1988). We are therefore obliged to use the less attractive torque approach, by evaluating stresses at the core–mantle boundary to the best of our ability.

Contributions to the time-series $\omega_i(t) = \Omega(m_1(t), m_2(t), 1+m_3(t))$ due to each of a wide variety of geophysical processes involved differ in their respective magnitudes and spectral characteristics. This together with the availability of daily values of $h_i^{(a)}(t)$ since the late 1970s is facilitating the determination of the slowly varying contribution

$$\tilde{\omega}_i = \Omega(\tilde{m}_1, \tilde{m}_2, 1+\tilde{m}_3) \tag{3.33}$$

to $\omega_i(t)$, which is due largely to the action of torques at the core–mantle interface. By equation (3.20) we find

$$\Omega C^{(m)} \dot{\tilde{m}}_3 + \Omega[\Delta \dot{I}_{33}^{(m,\,m)} + \Delta \dot{I}_{33}^{(m,\,c)}] = L_3^*, \tag{3.34}$$

where

$$L_3^* \equiv -(\Omega C^{(c)} \dot{\tilde{m}}_3 + \Omega \Delta \dot{I}_{33}^{(c)} + \dot{h}_3^{(c)}), \tag{3.35}$$

L_3^* being the axial component of the torque L_i^* produced on the mantle by the core. The quantity $\Delta I_{33}^{(m,\,m)}$ is the $(i = 3, j = 3)$ component of the change of $\Delta I_{33}^{(m)}$ associated with processes within the solid Earth (e.g. earthquakes, melting of ice), and $\Delta I_{33}^{(m,\,c)}$ is the $(i = 3, j = 3)$ component of the contribution to $\Delta I_{ij}^{(m)}$ resulting from the deformation of the solid Earth produced by the forces responsible for core–mantle coupling. Such deformations might be important (cf. Hinderer *et al.* 1987; Merriam 1988), but to the accuracy of any direct determination of these deformations and of the core–mantle torque L_i^* we are likely to be able to make in the foreseeable future, they represent a correction no greater than the uncertainties in L_i^*. Thus we write

$$\dot{\tilde{m}}_3 \approx L_3^* / \Omega C^{(m)} \tag{3.36}$$

as the leading approximation to the axial component of the equation of torque balance when 'atmospheric' and 'tidal' effects have been allowed for, at times when effects associated with $\Delta I_{ij}^{(m,\,m)}$ are negligible. This can be related to corresponding length-of-day changes through equation (3.8).

By equations (3.18) and (3.19), the leading approximations to the corresponding equatorial components of torque balance are the following:

$$A^{(m)}\Omega[\dot{\tilde{m}}_1 + \Omega A^{-1} \tilde{m}_2 (C-A)] = L_1^*, \tag{3.37a}$$

$$A^{(m)}\Omega[\dot{\tilde{m}}_2 - \Omega A^{-1} \tilde{m}_1 (C-A)] = L_2^*. \tag{3.37b}$$

When dealing with changes in $\tilde{m}_1$ and $\tilde{m}_2$ on timescales much greater than about a year, the terms involving $\dot{\tilde{m}}_1$ and $\dot{\tilde{m}}_2$ in the last two equations can be neglected, giving

$$(\tilde{m}_1, \tilde{m}_2) = [A/\Omega^2 (C-A)\, A^{(m)}]\,(-L_2^*, L_1^*). \tag{3.38}$$

4. Concluding remarks

The basic theoretical relation needed in the study of Earth rotation changes due to topographic torques at the core–mantle interface are given by equations (2.9), (3.36) and (3.38) or (3.37). The integrals on the right-hand side of equation (2.9) involve the core–mantle interface topography $h(\theta, \phi)$. When dealing with $L_1(t)$ and $L_2(t)$ and the polar motion $\tilde{m}_1(t)$ and $\tilde{m}_2(t)$ they produce, the dominant contribution to h is the equatorial bulge of the core–mantle interface, which corresponds to a 10 km difference between the equatorial radius and polar radius of the core. But the equatorial bulge makes no contribution to $L_3(t)$, which produces changes in the length of the day, so when dealing with such changes it is necessary to look in detail at features of h that depend on ϕ as well as θ. Over the past 20 years various attempts have been made to infer $h(\theta, \phi)$ from the pattern of long-wavelength gravity anomalies, with the most recent models incorporating the findings of seismic tomography and modern ideas about lower mantle rheology and convection (for references see papers by Hager & Richards and Woodhouse & Dziewonski, this Symposium).

The other quantity required in the evaluation of L_i from geophysical data is the field of horizontal velocity $\boldsymbol{u}_s = \boldsymbol{u}_s(\theta, \phi, t) = (0, v_s, w_s)$ in the free stream just below the core–mantle interface. Geomagnetic secular variation data have been used by various workers to infer $\boldsymbol{u}_s$ by a method based on the assumptions that: (*a*) on timescales very much less than that of free Ohmic decay of magnetic fields in the core, which is several thousand years for global-scale features, Alfvén's 'frozen magnetic flux' theorem can be applied, (*b*) in the outer reaches of the core the horizontal components of Coriolis forces are in geostrophic balance with the horizontal pressure gradient (see §2 above), and (*c*) the electrical conductivity of the mantle is everywhere very much less than that of the core; (for references see Whaler 1986; Jacobs 1987; Voorhies 1987; Bloxham 1988; Courtillot & Le Mouël 1988).

When Alfvén's theorem holds, the magnetic field $\boldsymbol{B}$ and the eulerian flow velocity $\boldsymbol{u}$ are related as follows:

$$\partial \boldsymbol{B}/\partial t = \nabla \times (\boldsymbol{u} \times \boldsymbol{B}). \tag{4.1}$$

Lines of force associated with long-wavelength features of the main geomagnetic field are advected by the horizontal flow $\boldsymbol{u}_s$ just below the core–mantle interface. Equation (4.1) above does not permit the unique determination of $\boldsymbol{u}_s$, but when the equation is combined with the geostrophic approximation expressed by equation (2.7) (see Backus & Le Mouël 1985; Voorhies 1987) it is possible in principle to determine $\boldsymbol{u}_s$ over most of the core, using as basic data values of $\boldsymbol{B}$ and $\partial \boldsymbol{B}/\partial t$ at the core–mantle interface, as obtained by the extrapolation of geomagnetic observations made at and near the Earth's surface. Various groups of geomagnetic workers are investigating the errors and uncertainties in the velocity fields so produced, which stem from imperfections in our knowledge of the spatial and temporal variations in the Earth's magnetic field and of the ranges of validity of the physical assumptions upon which the method is based.

A study of the application of equations (2.9), (3.36) and (3.38) making use of geophysical data being carried out in collaboration with R. W. Clayton, B. H. Hager, M. A. Spieth and C. V. Voorhies will be reported in a paper currently in preparation. The first results are encouraging, for they indicate that topographic coupling alone might account for the observed recent 'decade' changes in the Earth's rate of rotation without the necessity of invoking extreme models of $h(\theta, \phi)$ and $\boldsymbol{u}_s(\theta, \phi, t)$. Specifically, ϕ-variations in effective topographic height h of up to no more than about 0.5 km are implied by these calculations of topographic coupling. It might be significant that 0.5 km is also the magnitude of the departure from the equilibrium value of the equatorial bulge that Gwinn *et al.* (1986) (see also Wahr 1988) have inferred from their determinations of the amplitude and phase of free-core nutation based on very long baseline interferometry (VLBI) data. But 0.5 km is less by a factor of about 10 than the heights of irregular topography now being proposed by various workers on the basis of seismic tomography. This apparent discrepancy cannot be considered in detail here, but we note that the *effective* topographic height obtained by the method described in this paper should be the same as the actual height if the metallic core is in direct contact with the lower mantle. However, if there is a stable layer of poorly conducting low-viscosity liquid slag separating the metallic core from the solid mantle (D. L. Anderson, personal communication), then the pressure field acting on the actual topography could have weaker horizontal gradients than those present at the top of the metallic region, and the *effective* topographic height would in consequence be less than the actual height.

All of us involved with the investigation of the very difficult problems of determining the structure, composition and dynamics of the Earth's deep interior wish to see the development of suitable strategies for combining the wide variety of data available in the best possible way. It is to this end that the incomplete but in some ways novel material outlined in this paper is offered as a contribution to this Discussion Meeting on seismic tomography and mantle circulation.

References

Anufriev, A. P. & Braginski, S. I. 1977 Influences of irregularities in the boundary of the Earth's core on fluid velocity and magnetic field. *Geomagn. Aeron.* **17**, 78–82; 492–496.

Backus, G. E. & Le Mouël, J. L. 1987 The region on the core–mantle boundary where a geostrophic velocity field can be determined from the frozen-flux data. *Geophys. Jl R. astr. Soc.* **88**, 321–322.

Bloxham, J. & Gubbins, D. 1987 Thermal core–mantle interactions. *Nature, Lond.* **325**, 511–513.

Bloxham, J. 1988 The dynamical regime of fluid flow at the core surface. *Geophys. Res. Lett.* **15**, 585–588.

Bullard, E. C., Freedman, C., Gellman, H. & Nixon, J. 1950 The westward drift of the Earth's magnetic field. *Phil. Trans. R. Soc. Lond.* A **243**, 67–92.

Cazenave, A. (ed.) 1986 *Earth rotation: solved and unsolved problems.* (330 pages.) Dordrecht: D. Reidel Publishing Co.

Chao, B. F. & Gross, R. S. 1987 Changes in the Earth's rotation and low-degree gravitational field induced by earthquakes. *Geophys. Jl R. astr. Soc.* **91**, 569–596.

Courtillot, V. & Le Mouël, J.-L. 1988 Time variations of the Earth's magnetic field. *Ann. Rev. Earth planet. Sci.* **16**, 389–446.

Dickey, J. O., Eubanks, T. M. & Steppe, J. A. 1986 High accuracy Earth rotation and atmospheric angular momentum measurements. In *Earth rotation: solved and unsolved problems* (ed. A. Cazenave), pp. 137–162. Dordrecht: D. Reidel Publishing Co.

Doornbos, D. J. 1988 Multiple scattering by topographic relief with application to the core–mantle boundary. *Geophys. J.* **92**, 465–478.

Eltayeb, I. A. & Hassan, M. H. A. 1979 On the effects of a bumpy core–mantle interface. *Physics Earth planet. Inter.* **19**, 239–254.

Greenspan, H. P. 1968 *The theory of rotating fluids.* (327 pages.) Cambridge University Press.

Gwinn, C. R., Herring, T. A. & Shapiro, I. I. 1986 Geodesy by radio interferometry: studies of forced nutations of the Earth 2. Interpretation. *J. geophys. Res.* **91**, 4755–4765.

Hide, R. 1967 Motions of the Earth's core and mantle and variations of the main geomagnetic field. *Science, Wash.* **157**, 55–56.

Hide, R. 1969 Interaction between the Earth's liquid core and solid mantle. *Nature, Lond.* **222**, 1055–1056.

Hide, R. 1970 On the Earth's core–mantle interface. *Q. Jl R. met. Soc.* **96**, 379–390.

Hide, R. 1977 Towards a theory of irregular variations in the length of the day and core–mantle coupling. *Phil. Trans. R. Soc. Lond.* A **284**, 547–554.

Hide, R. 1984 Rotation of the atmospheres of the Earth and planets. *Phil. Trans. R. Soc. Lond.* A **313**, 107–121.

Hide, R. 1985 On the excitation of short-term variations in the length of the day and polar motion. *Geophys. Surv.* **7**, 163–167.

Hide, R. 1986 Presidential address – The Earth's differential rotation. *Q. Jl R. astron. Soc.* **278**, 3–14.

Hide, R. & Horai, K. I. 1968 On the topography of the core–mantle interface. *Physics Earth planet. Inter.* **1**, 305–308.

Hide, R. & Malin, S. R. C. 1971 Novel correlations between global features of the Earth's gravitational and magnetic fields: further statistical considerations. *Nature, Lond.* **230**, 63.

Hide, R. & Roberts, P. H. 1978 How strong is the magnetic field in the Earth's liquid core? *Physics Earth planet. Inter.* **20**, 124–126.

Hinderer, J., Legros, H., Gire, C. & Le Mouël, J.-L. 1987 Geomagnetic secular variation, core motions and implications for the Earth's wobbles. *Physics Earth planet. Inter.* **49**, 121–132.

Jacobs, J. A. 1975 *The Earth's core.* (253 pages.) London and New York: Academic Press.

Jacobs, J. A. 1984 *Reversals of the Earth's magnetic field.* (230 pages.) Bristol: Adam Hilger Ltd.

Jacobs, J. A. (ed.) 1987 *Geomagnetism* (2 volumes). New York: Academic Press Ltd.

Jault, D., Gire, C. & Le Mouël, J.-L. 1988 Westward drift, core motions and the exchange of angular momentum between core and mantle. *Nature, Lond.* **333**, 353–356.

Kalnay, E. & Mo, K. C. 1986 Mechanistic experiments to determine the origin of short-scale Southern Hemisphere stationary Rossby waves. In *Anomalous atmospheric flows and blocking* (ed. R. Benzi, B. Sultzman & A. C. Wiin-Nielsen). *Advances in geophysics, vol.* 29, pp, 415–442. Academic Press.

Lambeck, K. 1980 *The Earth's variable rotation.* (449 pages.) Cambridge University Press.

Le Mouël, J.-L. 1984 Outer-core geostrophic flow and secular variation of the geomagnetic field. *Nature, Lond.* **311**, 734, 735.

Malin, S. R. C. & Hide, R. 1982 Bumps on the core–mantle boundary: geomagnetic and gravitational evidence re-visited. *Phil. Trans. R. Soc. Lond.* A **306**, 281–289.

Melchior, P. 1986 *The physics of the Earth's core.* (256 pages.) Oxford: Pergamon Press.

Merriam, J. B. 1988 Limits on lateral pressure gradients in the outer core from geodetic observations. *Physics Earth planet. Inter.* **50**, 280–290.

Merrill, R. T. & McElhinny, M. W. 1983 *The Earth's magnetic field.* (395 pages.) New York and London: Academic Press Ltd.

Moffatt, H. K. 1978*a* *Magnetic field generation by fluid motion.* (343 pages.) Cambridge University Press.

Moffatt, H. K. 1978*b* Topographic coupling at the core–mantle interface. *Geophys. Astrophys. Fluid Dyn.* **9**, 279–288.

Moritz, H. & Mueller, I. I. 1987 *Earth rotation: theory and observation.* (617 pages.) New York: The Ungar Publishing Company.

Munk, W. H. & Macdonald, G. J. F. 1960 *The rotation of the Earth.* (323 pages.) Cambridge University Press.

Paulus, M. & Stix, M. 1986 Electromagnetic core–mantle coupling. In *Earth rotation solved and unsolved problems* (ed. A. Cazenave), pp. 259–267. Dordrecht: D. Reidel Publishing Co.

Roberts, P. H. 1989 Core–mantle coupling. *Encyclopedia of geophysics* (ed. D. E. James). Stroundsburg, Pennsylvania: Van Nostrand.

Rochester, M. G. 1984 Causes of fluctuations in the Earth's rotation. *Phil. Trans. R. Soc. Lond.* A **313**, 95–105.

Runcorn, S. K., Creer, K. M. & Jacobs, J. A. (ed.) 1982 The Earth's core: its structure, evolution and magnetic field. *Phil. Trans. R. Soc. Lond.* A **306**, 1–289.

Silver, P. G. & Carlson, R. W. 1988 Deep slabs, geochemical heterogeneity, and the large-scale structure of mantle convection. *Ann. Rev. Earth planet. Sci.* **16**, 477–541.

Vogel, A. 1960 Über Unregelmässigkeiten der äusseren Begrenzung des Erdkerns auf Grund von am Erdkern reflektierten Erdbebenwellen. *Gerlands Beiträge zur Geophysik* **69** (3), 150–174.

Vondrák, J. & Pejović, N. 1988 Atmospheric excitation of polar motion: Comparison of the polar motion spectrum with spectra of effective atmospheric angular momentum functions. *Bull. astr. Inst. Csl* **39**, 172–185.

Voorhies, C. V. 1987 The time-varying geomagnetic field. *Rev. Geophys.* **25**, 929–938.

Wahr, J. M. 1988 The Earth's rotation. *Ann. Rev. Earth planet. Sci.* **16**, 231–249.

Whaler, K. 1986 Geomagnetic evidence for fluid upwelling at the core–mantle boundary. *Geophys. Jl R. astr. Soc.* **86**, 563–588.

Phil. Trans. R. Soc. Lond. A **328**, 365–375 (1989)
Printed in Great Britain

Implications of geomagnetism for mantle structure

By D. Gubbins

Department of Earth Sciences, University of Cambridge, Bullard Laboratories, Madingley Rise, Madingley Road, Cambridge CB3 0EZ, U.K.

Recent studies of the magnetic field at the core–mantle boundary have revealed fixed sites of either static magnetic features or persistent secular variation. This suggests that part of the magnetic-field behaviour is controlled by the mantle. The most plausible mechanism for core–mantle interaction is thermal coupling, although topography may also be significant. The magnetic sites coincide with anomalies in lower-mantle seismic velocity, as determined from tomography, and density, as determined by flow models of mantle convection constrained by tomography and the geoid. Some magnetic features coincide with subduction zones, particularly those near Indonesia; they may be caused by bumps on the core–mantle boundary beneath trenches. Palaeomagnetic pole positions suggest the magnetic behaviour has persisted for at least 5 Ma, as would be expected if it were controlled from the mantle. These conclusions could be quantified if the frozen-flux hypothesis allowed determination of fluid flow at the core surface, but unfortunately failure of the hypothesis makes all such determinations suspect. Core motions calculated so far suggest the flow is mainly toroidal. Questions about the dynamics of the flow (whether it is steady, stratified, or geostrophic) remain unresolved.

1. Introduction

Secular variation (sv) is the slow change in the Earth's magnetic field that is caused by fluid motion in the liquid core, rather than by external effects. These fluid flows may be an integral part of the dynamo process responsible for generation of the main field, in which case an understanding of sv would lead to a better understanding of the dynamo process, or they may be quite separate (operating on different time scale and in a different part of the liquid core: near the core–mantle boundary, CMB, with the dynamo operating deeper down, for example).

The past five years has seen considerable advances in our understanding of sv. First, a sudden change in 1970 at many observatories in Europe and elsewhere, now called the *jerk*, showed that core fields change on a more rapid timescale than had hitherto been suspected (a review is given in Courtillot & LeMouël 1984). The changes are probably associated with changes in the length of day (Vestine 1952; Le Mouël *et al.* 1981). Secondly, the satellite, MAGSAT (Langel *et al.* 1980) provided the best 'snapshot' of the magnetic field to date and stimulated collaborative research on sv. Thirdly, new studies of old magnetic observations (Gubbins & Bloxham 1985; Bloxham & Gubbins 1985, 1986; Langel *et al.* 1986; Bloxham *et al.* 1989), analysed by new inverse theoretical techniques (Whaler & Gubbins 1980; Shure *et al.* 1982; Parker & Shure 1982; Gubbins 1983; Gubbins & Bloxham 1985), has provided a much better picture of the sv during the past three centuries than had previously been available. There have been several attempts to determine fluid flow in the core from sv, which are described in §3.

The radial component of magnetic field at the CMB for epoch 1980 is shown in figure 1.

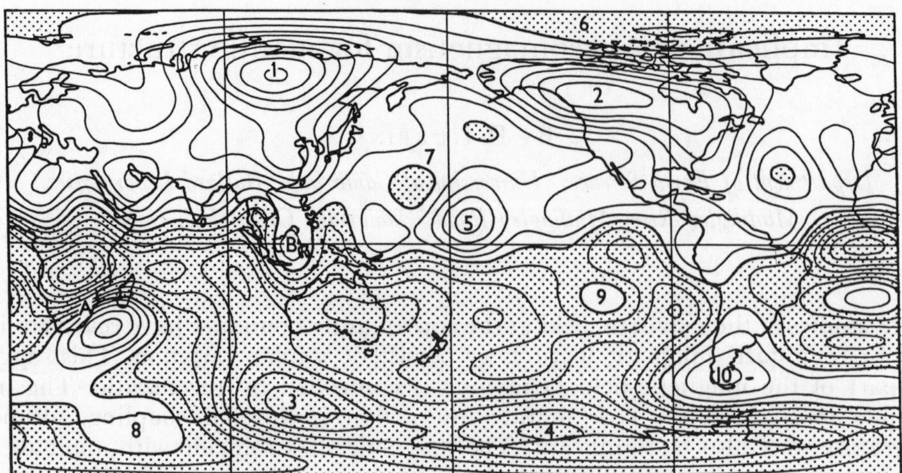

FIGURE 1. Radial component of magnetic field at the CMB, 1980 (after Gubbins & Bloxham 1985). Contour interval is 100 μT. The thick lines are of zero-radial component: the *null-flux curves*. Features 1–9 have remained nearly stationary throughout the past 300 years. Most of the Earth's dipole moment is contained in features 1–4, called the *main lobes*. Feature 10 and those near A have drifted rapidly westwards. B is the site of rapid *in situ* oscillations.

This model of the main magnetic field is based on MAGSAT satellite data. It was derived by Gubbins & Bloxham (1985), but any model based on a spherical harmonic expansion truncated or tapered near degree 14 will produce a similar map. The analysis assumes the mantle is an electrical insulator. Bloxham *et al.* (1989) have also derived a sequence of field models for epochs 1715, 1777, 1842, 1888, 1905, 1915, 1925, 1935, 1945, 1955, 1966, and 1969, based on nearly 180000 magnetic measurements and using a consistent method of analysis that produces smooth core fields. This 285 year long record gives an excellent picture of recent SV. The time span is long enough to follow movement of certain core-field features across many degrees of longitude, and to observe other features that have remained stationary. The main feature of core-field SV is the westward drift, as it is for surface fields. However, the drift is not global, and is almost completely absent from the Pacific region. One region of low radial flux near Easter Island (feature 9 in figure 1) remains stationary, whereas another patch of flux (near A in figure 1) can be followed from near B in figure 1 in 1777 to its present position near Africa in the same interval of time.

This combination of drifting and fixed features in the field led Bloxham & Gubbins (1987) to suggest the SV was controlled by thermal core–mantle interactions. In this paper, I argue that, as a consequence of core–mantle interactions, published geomagnetic and paleomagnetic data can provide valuable clues to the thermal state and structure of the lower mantle and core. If geomagnetism can provide independent confirmation of proposed models for the lower mantle obtained by seismological and mantle-convection studies, it gives us hope that one day the lower mantle will be mapped and understood in as much detail as the lithosphere. The present situation is similar to that in plate tectonics in 1960. There was then a great deal of circumstantial evidence for contintental drift, some of it dating back several decades. The evidence provided by geomagnetism, through the reversal timescale, converted this circumstantial evidence into a quantitative theory.

2. Frozen-flux theory and determination of fluid flow at core surface

Roberts & Scott (1965) assumed the core fluid was a perfect conductor and therefore the field lines were frozen to the fluid, by Alfvén's theorem, and could be used as tracers for the flow. Kahle *et al.* (1967) attempted a calculation of fluid flow, but there followed a lull when it was realized that determinations of field at the CMB were poor and determinations of fluid flow from Roberts & Scott's frozen-flux hypothesis were highly non-unique, even with perfect data. Backus (1968) showed that *null-flux curves*, lines where the magnetic field at the core surface is horizontal, are material lines and move with the fluid. It is possible to determine fluid flow normal to these curves but not along them. He also showed the flux through patches of the CMB bounded by null-flux curves remains invariant. Null-flux curves are shown as thick solid lines in figure 1.

Booker (1969) attempted to test the frozen-flux hypothesis by calculating the flux through the northern magnetic hemisphere for several field models. He observed a change associated with the fall in dipole moment, but did not regard it as a violation of the hypothesis because flux could have passed into small-scale features not resolved by his field models. Gubbins & Bloxham (1985) recalculated the flux through the northern magnetic hemisphere using later field models and found in favour of the hypothesis. However, they subsequently studied smaller patches and found a significant change in flux in the south Atlantic region (Bloxham & Gubbins 1986) between epochs 1960, 1970 and 1980, which violated the hypothesis. This small change has now been confirmed as part of a long-term trend dating from the beginning of the century (Bloxham & Gubbins 1985; Bloxham *et al.* 1988). It is not a particularly surprising result, because typical estimates of the electrical conductivity of iron in the core (5×10^5 S m^{-1}) suggest diffusion times of about 100 years for features of the size considered. The frozen-flux hypothesis appears to hold away from the south Atlantic region.

Backus (1968) also derived conditions on the horizontal component of field for the frozen-flux hypothesis to hold. The horizontal components cannot be used to resolve the non-uniqueness in determining fluid flow except on the null-flux curves, although they do provide some information about the shear. In a perfectly conducting core a current sheet may form in the boundary layer at the top of the core, and horizontal components of field may be discontinuous across the CMB. Surface observations of magnetic field only allow determination of the field at the base of the mantle and will not therefore be usable if the jump across the boundary layer is large. Roberts & Scott (1965) analysed the boundary layer and concluded that the jump was negligible; Backus (1968) reanalysed it but came to no firm conclusion; whereas Hide & Stewartson (1972) included Coriolis forces and again concluded the jump was negligible. Barraclough *et al.* (1989) have calculated Backus's conditions and found them to hold for epochs 1960–1980. Thus if we allow use of the radial field to determine core flows, it is consistent to use the horizontal components also.

Determination of fluid flow from magnetic observations requires some regularization to prevent errors in the data from mapping into spurious fluid flows. This regularization also removes the fundamental ambiguity that is present even with perfect data. It is therefore an *ad hoc* means of selecting one particular core flow that fits the data. Several models of fluid flow have been produced in this way, after the pioneering work of Kahle *et al.* (1967), and all suffer from the same fault. More recently, attempts have been made to use the dynamical equations to remove the ambiguity. Three assumptions have been made. I shall discuss each in turn.

First, the flow was assumed stratified and therefore toroidal (Gubbins 1982). The reasons were:

(1) that the heat flow required to sustain the dynamo is rather high and it may not be possible to exceed the adiabatic gradient everywhere in the core; the adiabatic gradient steepens near the CMB, because of the increase in gravitational acceleration across the core, and therefore the most likely site for a sub-adiabatic gradient and stable fluid is near the CMB (Gubbins *et al.* 1982);

(2) low SV occurs near stationary points in the core field, a consequence of stratified flow (Whaler 1980).

Toroidal flows preserve the flux through every patch of the CMB bounded by a contour of radial field (not simply the null-flux curves). They allow determination of fluid flow normal to the contours (but not along them). The ambiguity in determining fluid motion is alleviated but not completely removed by assuming toridal motion. The hypothesis can be tested by calculating the flux integrals that should be conserved, but this has not been done. Several authors have carried out calculations of full fluid velocities and claim the poloidal part is significantly greater than zero, thus contradicting the toroidal hypothesis (Whaler 1986; Gire *et al.* 1986; Voorhies 1986). These statistical arguments are unconvincing, however: they take no account of departures from frozen flux, which are likely to produce spurious poloidal motions.

Secondly, steady flows have been assumed. If the Lorentz forces are weak, the timescale for changes in the fluid velocity might be long compared with that for changes in the magnetic field. In practice, some form of steady flow is always assumed, because SV is estimated by differencing field models at separate epochs. The hypothesis allows unique determination of fluid motion, which is its main attraction. Voorhies & Backus (1985) have proved that any steady flow can be determined uniquely from field models from at least three different epochs. The third epoch allows a second determination of core motion that, provided the field contours are different, allows the non-uniqueness to be resolved. In practice, sufficient time must elapse between the two determinations for the contours to change significantly. Gubbins (1984) studied the 'jerk' interval around 1970 and concluded that the data were inconsistent with steady motion. Nevertheless, there have been several determinations of steady fluid motion (Voorhies 1986; Whaler & Clarke 1988; Bloxham 1988). These authors claim the steady-motion hypothesis is satisfied, in contradiction to Gubbins's result. This may be because they have excluded the jerk or omitted the fine time resolution afforded by observatory annual means around the time of the jerk.

The constraint of steady flow takes the form of a vanishing determinant (Voorhies & Backus 1985). If insufficient time has elapsed between epochs the constraint will be ineffective in restricting the class of fluid flows satisfying the data. It is very likely, therefore, that published 'unique' steady motions are unique because of the regularizing condition used to suppress noise in the determination, and not because of the steady-motion requirement. They are therefore no less *ad hoc* than determinations made without the steady-motion condition. Increasing the time interval, or the accuracy of the field models, is likely to introduce errors due to failure of the frozen-flux hypothesis. The theorem is therefore likely to be of limited use.

Thirdly, LeMouël *et al.* (1985) have assumed tangentially geostrophic flow. This requires the

Lorentz force to be absent from the radial component of the vorticity equation near the CMB. The flow can be determined uniquely within patches of the CMB enclosed by null-flux curves that intersect the geographic equator (see also Backus & LeMouël 1986). Geostrophic motions have been calculated by Gire *et al.* (1986) and Bloxham (1988). The geostrophic constraint takes the form of difference equations that must be satisfied by the spherical harmonic coefficients of the velocity Gire *et al.* (1986). There is a truncation problem: if the velocity series is truncated the geostrophic constraint will not be satisfied by the highest-order harmonics. If the constraint is satisfied perfectly then the high-order harmonices become distorted and it is difficult to fit the observations satisfactorily. Gire *et al.* (1986) applied the geostrophic constraint approximately, in the hope that high harmonics are unimportant, but the truncation problem appears to be very severe. The geostrophic and toroidal motion constraints are $\nabla_h \cdot (\boldsymbol{v} \cos \theta) = 0$ and $\nabla_h \cdot \boldsymbol{v} = 0$ respectively. I have calculated both quantities and found them to be of comparable magnitude for the motions of Gire *et al.* (1986). Bloxham (1988) has calculated both toroidal and fully geostrophic velocities and finds toroidal motions fit the field models better than geostrophic motions. LeMouël (personal communication 1988) has now found satisfactory flows that are fully geostrophic. Thus no calculation demonstrates the core flow satisfies either geostrophic or stratified dynamics.

Maps of toroidal, geostrophic, and steady velocities all account for westward drift of the field pattern in the equatorial Atlantic region and absence of SV in the Pacific. The fluid flow is predominantly westward in the Atlantic hemisphere and absent from the Pacific. The geostrophic and unconstrained flows have upwelling in the Indian Ocean and downwelling off the coast of Peru to allow for this, whereas the toroidal flows return eastwards at high latitudes, avoiding the quiescent Pacific region (for example, see the flows in LeMouël *et al.* 1985). I believe the poloidal part of the flow is rather small and is not adequately determined by any of these calculations; those places where upwelling appears are regions where the frozen-flux hypothesis is suspect because of flux expulsion (Bloxham 1986), and downwelling appears off Peru simply because it must occur somewhere in order to conserve mass.

Core velocity calculations are instructive in providing maps of fluid flow that may aid our intuition, but they are not useful in establishing the dynamical régime in a rigorous statistical manner; in fact, the results may be downright misleading. The flow may appear toroidal (or geostrophic) when in reality it is not. For example, plate movements on the Earth's surface appear horizontal almost everywhere: upwelling and downwelling is confined to ridges and trenches, which occupy very little area. Core convection may be similar, which could explain Whaler's (1980) result. If such is the case, toroidal motions may approximate the flow quite well in most places, but the underlying hypothesis would be quite false.

3. Thermal core–mantle interactions

Bloxham & Gubbins (1985) suggested several sites of stationary field or of persistent SV activity were tied to the solid mantle. In figure 1, features 6, 7, 8 and 9 are stationary and have low flux; 1–5 are stationary regions of high flux; A is the site of apparent expulsion of toroidal flux; and the undulation in the magnetic equator near B is a site of persistent *in situ* oscillations. Flux concentration could result from fluid downwelling (by Alfvén's theorem the field lines are swept towards the point of downwelling); low flux and toroidal flux expulsion from upwelling. Persistent *in situ* oscillations could be a consequence of a topographic feature on the CMB.

Three mechanisms could explain the mantle's influence of SV. Lateral variation of electrical conductivity in the mantle could preferentially shield SV in, for example, the Pacific region, where it is low. Extreme variations in conductivity are required. The idea is not considered further here. The other two mechanisms are thermal interaction, in which temperature variations in the boundary layer at the base of the mantle drive thermal winds in the core, as proposed by Bloxham & Gubbins (1987), and topography on the CMB, as proposed by Hide (1967) (see also Gubbins & Richards 1986).

Consider thermal interactions first. Lower-mantle convection is likely to be accompanied by lateral variations in temperature of several tens of degrees. There is evidence for lateral variations in seismic velocity in the lower mantle, which could be accounted for by such temperature variations. The core fluid cannot withstand any sizeable lateral variation in temperature, and differences of as much as one degree will drive thermal winds that tend to restore equipotential surfaces to uniform temperature. The mantle will therefore experience a constant temperature, variable heat flux, lower boundary.

There is no doubt that mantle convection will entrain core motions; the only question is whether heat flux variations across the CMB are large enough to drive sufficiently strong core flows to influence the SV: about 0.1–1.0 mm s^{-1}. These flows must carry heat from regions on the CMB where heat transfer into the mantle is low to regions where it is high. The heat carried by the core fluid can be estimated in terms of the fluid velocity, the thermal-diffusion skin depth, and the very small temperature drop experienced by the core fluid in its passage along the CMB (Bloxham & Gubbins 1987). The flow speed can also be related to the small temperature drop by using the thermal wind equation. The resulting heat transfer is of the same order of magnitude as the lateral variation in heat flux into the mantle expected on the basis of mantle convection studies: as much as 100% of the average heat flux (G. A. Houseman, personal communication 1987). Thermal interactions are therefore a plausible agent for sustaining core flows responsible for the SV.

Core fluid must flow from regions of low heat flux, across the CMB, to regions of high heat flux; therefore in the simplest case we expect upwelling beneath the former and downwelling beneath the latter. Regions of high heat flow correspond to low temperatures in the thermal boundary layer at the base of the mantle, and consequently fast seismic velocity (assuming the lower mantle is laterally chemically homogeneous). Bloxham & Gubbins (1987) therefore proposed upwelling of core fluid where the mantle boundary layer was hot, and downwelling where it was cold.

This mechanism for thermal core–mantle interaction can be made quantitative. Assume the frozen-flux conditions hold and the jump in horizontal component of magnetic field across the CMB is negligible. Assume also that the flow is geostrophic, and that the Lorentz force is absent from all three components of the vorticity equation. This set of assumptions goes beyond those made by LeMouël *et al.* (1985), who used only the radial components of the vorticity and induction equations. The vorticity equation is

$$(\boldsymbol{\Omega}\cdot\nabla)\boldsymbol{v} = -\alpha\nabla T\times\boldsymbol{g}, \qquad (1)$$

where α is the thermal expansion coefficient and T is the (nearly uniform) temperature at the CMB. This equation, plus the full magnetic induction equation, allows determination of T everywhere, provided the magnetic field satisfies a certain point condition on the CMB.

All published velocity models show predominantly westward flow in the Atlantic hemisphere.

Equation (1) leads to westward flow if the temperature gradient is in a north–south direction. The ϕ component of (1) gives

$$\frac{\partial v_\phi}{\partial z} = \frac{\alpha g}{\Omega r}\frac{\partial T}{\partial \theta}, \tag{2}$$

which gives a maximum shear at the Equator for a dipolar temperature distribution. This dipolar form can dominate the temperature distribution when a formal inversion is carried out (work in progress). This is a very unsatisfactory result, because the westward drift is attributed to a specific CMB temperature distribution: *one for which there is no independent corroborative evidence.* The westward drift is more likely the consequence of magnetohydrodynamic instabilities, which tend to migrate westward (see, for example, Acheson & Hide 1973). Rather than accept such an implausible explanation for westward drift, we must abandon one of the hypotheses that leads to the determination of temperature. The frozen-flux hypothesis is known to fail in certain regions, but the westward drift is unlikely to be driven by diffusion. The more likely candidate is presence of a Lorentz force in equation (1). The Lorentz force contains derivatives of magnetic field that are not continuous across the CMB, and therefore it cannot be estimated from surface data. Further formal inversion is not possible, and an *a priori* model of Lorentz force is needed. This is an illustration as to how formal inversion can be misleading: the temperature profiles fit the data quite well, and all magnetic field maps satisfy the consistency conditions tolerably well, but the results are physically implausible.

Hide (1967) proposed that topography could influence core flows strongly. There are sound arguments for believing that topographic torquing of the mantle is responsible for observed changes in the length of day (Hide, this Symposium). This coupling is rather different from the longer-term interactions under discussion here. Hide & Malin (1970) proposed a correlation between the non-dipole field, rotated eastwards by about 140°, and the low spherical–harmonic degree terms in the geoid, to support the idea. In terms of the new CMB fields, such as those in figure 1, Hide & Malin's correlation (see also Malin & Hide 1982) appears to be mainly between the rapidly westward-drifting features, such as A in the Atlantic region, and a geoid anomaly almost coincident with the Pacific Basin. There seems no reason to select these two features for correlation, and it is very hard to understand how a topographic feature on the CMB can shed an image of itself in the core field, which subsequently drifts to the west. A bump is likely to produce standing oscillations and generate propagating instabilities or waves, as appears to happen near B of figure 1.

A bump also has a thermal effect. If we cause a bump to form on the CMB, so that mantle protrudes into the core, it will entrain convection in the core in exactly the same way as if we had made that part of the mantle cold. The bump introduces a cold temperature at depth within the core. It is therefore difficult to separate topographic and thermal anomalies in a convecting mantle. Cold, descending mantle material will cause both a bump and a cold patch on the CMB, so both effects are inextricably linked. It is possible, in principle, to determine CMB topography independently from the seismic velocity of the lower mantle, and thereby separate the two effects. If the mantle is chemically inhomogenous the effects of bumps and temperature will be difficult.

4. Confirming evidence

The theory of core–mantle interactions outlined in the previous section is predictive, and can therefore be confirmed or disproved by using different observations. If the sv is controlled by the mantle, then we should expect the same field morphology and pattern of sv to apply throughout the recent geological past. The historical record may be too short to provide a 'typical' example of field behaviour, but there is hope that palaeomagnetism can provide evidence of long-term geomagnetic changes. Palaeomagnetic pole determinations have been used to reconstruct past plate motions. In the past 5 Ma the plates have not moved significantly and the collection of pole positions allow study of the non-dipole field. One well-known feature of these pole positions is 'far-sidedness' (Wilson 1970), in which a palaeomagnetic pole always appears further away than the geographical pole. The effect persists throughout many reversals. I have suggested this is an effect of geographical distribution of the data (Gubbins 1988). The four main lobes of the field (1–4 in figure 1) cause far-sidedness in the extreme north Atlantic, as observed, but *near*-sidedness in the corresponding latitudes in the Pacific hemisphere; the reverse flux features (such as 10 in figure 1) cause *near*-sidedness in the south Atlantic. Significant *near*-sidedness is found in the south Atlantic and north Pacific regions when the palaeomagnetic pole data are grouped according to geographical location. Unfortunately there are rather few pole determinations from these areas. The model also predicts near-sidedness in the southeast Pacific, which is not observed. No other satisfactory explanation has been offered for the far-sided bias to the pole positions.

Seismic tomography gives an estimate of seismic velocity in the lower mantle that may be interpreted as temperature if the mantle is chemically homogeneous. A map of velocity anomalies derived by Dziewonski (1984) is shown in figure 2. The fast regions (marked +) coincide with flux lobes 1 (contour +125 in figure 2) and 2 (contour +75 in figure 2), and the

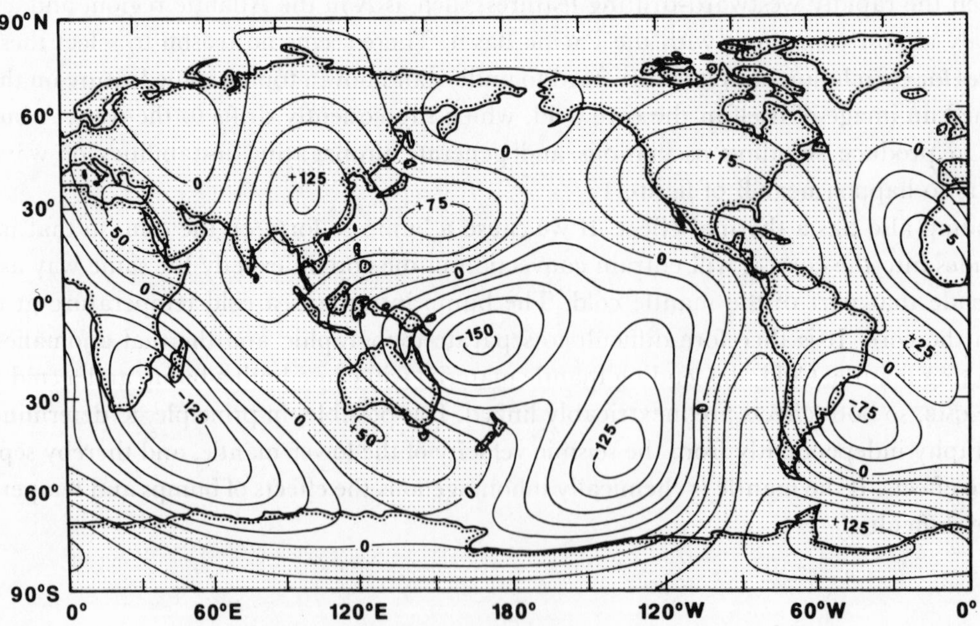

FIGURE 2. Lateral variations in P-wave velocity in the lowermost mantle, after Dziewonski (1984). Contour interval is 25 m s^{-1}.

slow region (contour -125 in figure 2) with the region of flux expulsion (A in figure 1). The agreement is rather qualitative, but is encouraging support for thermal interactions. The main core features are likely to be generated with antisymmetry about the equator (Gubbins & Bloxham 1987) and therefore if lobes 1 and 2 are tied, then 3 and 4 would be fixed. In fact we should not expect perfect coincidence of features. Gubbins & Bloxham (1987) argue that the flux lobes must lie close to the inner core circle if they are part of the main dynamo field and therefore a manifestation of deep-core convection; they may therefore be concentrated some way away from the temperature anomaly (such as appears to be the case with lobe 1).

A better and more surprising correlation is shown by a calculation of 'dynamical CMB topography' using the method of Hager *et al.* (1985). A model of mantle convection is based on the geoid and seismic tomography data assuming chemical homogeneity and a simple relation between density and seismic velocity. The CMB topography results from temperature variations throughout the lower mantle, and is therefore a good guide to the overall temperature anomaly in the lowermost mantle. Figure 3 illustrates the results of one such calculation; the expected temperature anomalies are all present, including hot regions near magnetic features 6, 7, 8 and 9, and a small cold spot near 5. It is not possible to do full justice to the comparison of the various geomagnetic and tomographic maps in this short review. A more complete discussion and a full set of figures is given in Bloxham *et al.* (1989).

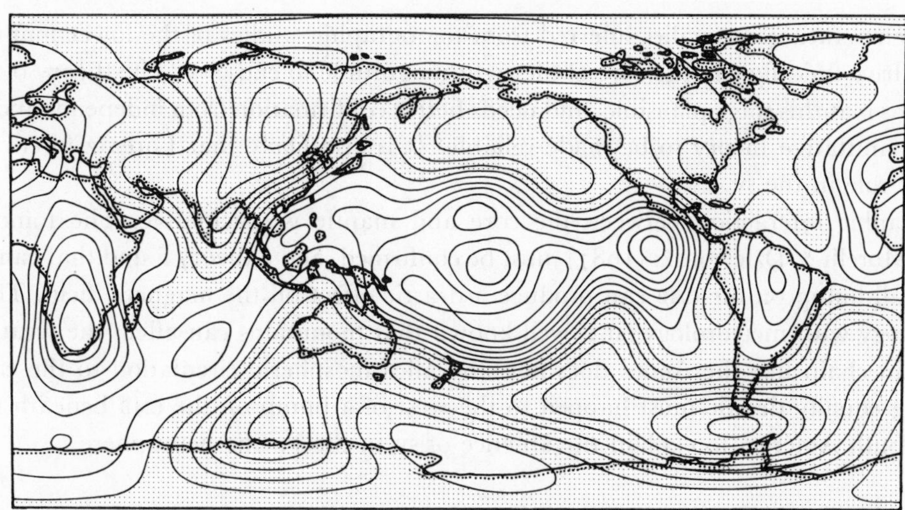

FIGURE 3. 'Dynamic CMB topography' as calculated by the method of Hager *et al.* (1985) from Gubbins & Richards (1986). Assumed viscosity models and relations between seismic velocity and density are constrained by the geoid and seismic tomography to produce density and temperature anomalies throughout the convecting mantle. Deformations to the interfaces are also produced, as shown here for the CMB. The contour interval is 400 m. These deformations are directly related to an average of the temperature variations in the overlying mantle.

Morelli & Dziewonski (1987) have determined a model of the CMB directly from observations of PcP and PKP waves. Surprisingly, they find significant topography beneath the subduction zones. It would be easy to dismiss this result as perhaps an earthquake source effect, but the geomagnetism lends some support to it. The region of very low SV in the Pacific is delineated well by the ring-of-fire around the Pacific: perhaps there is a connection via bumps on the CMB. The topography on the CMB is beneath the trenches and not at the base of subducted lithosphere; it must therefore be associated with a mass anomaly at the trench rather than some effect of the whole slab. This observation must remain questionable until the seismological result is confirmed and some satisfactory mechanism is found for topographic interaction between the core and mantle.

5. Conclusions

I have reviewed a number of new ideas for driving SV. The main conclusion is that formal inversions for fluid velocity at the core surface are likely to yield maps that will aid our intuition, but they cannot be used in a quantitative way to confirm or deny hypotheses for the underlying dynamics. Future advances are more likely to come from developing specific models (for westward-drifting instabilities, for example), and making qualitative comparisons with observations, than from inversion.

The qualitative ideas presented in this paper can be classified as reliable, plausible, or speculative (depending more on the nature of the theory than the statistical fit to the observations). Reliable results include the field morphology at the CMB, the general pattern of SV (such as westward drift in the Atlantic hemisphere and low SV in the Pacific region, oscillations and drift of persistent features in the field, even some quite small ones, and predominantly westward fluid motion in the Atlantic hemisphere region, a feature shared by all velocity determinations). It is also quite clear that flux diffusion occurs on parts of the CMB.

If the temperature anomalies inferred from tomographic studies of the lower mantle exist, they will drive fluid flow in the core of the right magnitude to cause SV. The theory of thermal interactions is at this stage plausible, and needs better confirmation from independent data and a more quantitative theory to be made reliable. It is also likely that Lorentz forces play an important role in the SV.

Topographic interaction between the core and mantle is speculative. The tomographic results of Morelli & Dziewonski (1987) must be confirmed, and a theory found that can explain how a small bump on the CMB can produce a drastic effect on the magnetic field. There is a good analogy with meteorology: it is well known that mountains can affect the weather, but a mountain chain like the Andes does not stop all atmospheric circulation to the east of it. Something equally drastic must happen to the west of a bump on the CMB beneath the west coast of South America to explain the absence of SV in the Pacific hemisphere.

References

Acheson, D. J. & Hide, R. 1973 *Rep. Prog. Phys.* **36**, 159–221.
Backus, G. E. 1968 *Phil. Trans. R. Soc. Lond.* A **262**, 239–266.
Backus, G. E. & LeMouël, J.-L. 1986 *Geophys. Jl R. astr. Soc.* **85**, 617–628.
Barraclough, D. R., Gubbins, D. & Kerridge, D. 1989 *Geophys. Jl R. astr. Soc.* (Submitted.)
Bloxham, J. 1986 *Geophys. Jl R. astr. Soc.* **87**, 669–678.
Bloxham, J. 1988 *Geophys. Res. Lett.* **15**, 585–588.
Bloxham, J. & Gubbins, D. 1985 *Nature, Lond.* **317**, 777–781.
Bloxham, J. & Gubbins, D. 1986 *Geophys. Jl R. astr. Soc.* **84**, 139–152.
Bloxham, J. & Gubbins, D. 1987 *Nature, Lond.* **325**, 511–513.
Bloxham, J., Gubbins, D. & Jackson, A. 1989 *Phil. Trans. R. Soc. Lond.* (Submitted.)
Booker, J. R. 1969 *Proc. R. Soc. Lond.* A **309**, 27–40.
Courtillot, V. & LeMouël, J.-L. 1984 *Nature, Lond.* **311**, 709–716.
Dziewonski, A. M. 1984 *J. geophys. Res.* **89**, 5929–5952.
Gire, C., LeMouël, J.-L. & Madden, T. 1986 *Geophys. Jl R. astr. Soc.* **84**, 1–29.
Gubbins, D. 1982 *Phil. Trans. R. Soc. Lond.* A **306**, 247–254.

Gubbins, D. 1983 *Geophys. Jl R. astr. Soc.* **73**, 641–652.
Gubbins, D. 1984 *Geophys. Jl R. astr. Soc.* **77**, 753–766.
Gubbins, D. 1988 *J. geophys. Res.* **93**, 3413–3420.
Gubbins, D. & Bloxham, J. 1985 *Geophys. Jl R. astr. Soc.* **80**, 696–713.
Gubbins, D. & Bloxham, J. 1987 *Nature, Lond.* **325**, 509–511.
Gubbins, D. & Richards, M. 1986 *Geophys. Res. Lett.* **13**, 1521–1524.
Gubbins, D., Thomson, C. J. & Whaler, K. A. 1982 *Geophys. Jl R. astr. Soc.* **68**, 241–251.
Hager, B. H., Clayton, R. W., Richard, M. A., Comer, R. P. & Dziewonski, A. M. 1985 *Nature, Lond.* **313**, 541–545.
Hide, R. 1967 *Science, Wash.* **157**, 55–57.
Hide, R. & Malin, S. R. C. 1970 *Nature, Lond,* **225**, 605–609.
Hide, R. & Stewartson, K. 1972 *Rev. Geophys. Space Phys.* **10**, 579–598.
Kahle, A. B., Vestine, E. H. & Ball, R. H. 1967 *J. geophys. Res.* **72**, 1095–1108.
Langel, R. L., Estes, R. H., Mead, G. D., Fabiano, E. B. & Lancaster, E. R. 1980 *Geophys. Res. Lett.* **7**, 793–796.
Langel, R. L., Kerridge, D. J., Barraclough, D. R. & Malin, S. R. C. 1986 *J. Geomagn. Geoelect, Kyoto.* **38**, 573–597.
LeMouël, J.-L., Gire, C. & Madden, T. 1985 *Physics Earth planet. Inter.* **39**, 270–287.
LeMouël, J.-L., Madden, T. R., Ducruix, J. & Courtillot, V. 1981 *Nature, Lond.* **290**, 763–765.
Malin, S. R. C. & Hide, R. 1982 *Phil. Trans. R. Soc. Lond.* A **306**, 281–289.
Morelli, A. & Dziewonski, A. M. 1987 *Nature, Lond.* **325**, 678–683.
Parker, R. L. & Shure, L. 1982 *Geophys. Res. Lett.* **9**, 812–815.
Roberts, P. H. & Scott, S. 1965 *J. Geomagn. Geoelect, Kyoto* **17**, 137–151.
Shure, L., Parker, R. L. & Backus, G. E. 1982 *Physics Earth planet. Inter.* **28**, 215–229.
Vestine, E. H. 1952 *Proc. natn. Acad. Sci. USA.* **38**, 1030–1038.
Voorhies, C. V. 1986 *J. geophys. Res.* **91**, 12444–12466.
Voorhies, C. V. & Backus, G. E. 1985 *Geophys. Astrophys. Fluid Dyn.* **32**, 163–173.
Whaler, K. A. 1980 *Nature, Lond.* **287**, 528–530.
Whaler, K. A. 1986 *Geophys. Jl R. astr. Soc.* **86**, 563–588.
Whaler, K. A. & Clarke, S. O. 1988 *Geophys. Jl R. astr. Soc.* **94**, 143–155.
Whaler, K. A. & Gubbins, D. 1980 *Geophys. Jl R. astr. Soc.* **65**, 645–693.
Wilson, R. L. 1970 *Geophys. Jl R. astr. Soc.* **19**, 417–437.

Phil. Trans. R. Soc. Lond. A **328**, 377–389 (1989)
Printed in Great Britain

Density and composition of the lower mantle

By R. Jeanloz and Elise Knittle

Department of Geology and Geophysics, University of California, Berkeley, California 94720, *U.S.A.*

The observed density distribution of the lower mantle is compared with density measurements of the $(Mg,Fe)SiO_3$ perovskite and (Mg,Fe)O magnesiowüstite high-pressure phases as functions of pressure, temperature and composition. We find that for plausible bounds on the composition of the upper mantle (ratio of magnesium to iron + magnesium components $x_{Mg} \geqslant 0.88$) and the temperature in the lower mantle ($T \geqslant 2000$ K), the high-pressure mineral assemblage of upper-mantle composition is at least 2.6 (± 1) % less dense than the lower mantle over the depth range 1000–2000 km. Thus, we find that a model of uniform mantle composition is incompatible with the existing mineralogical and geophysical data. Instead, we expect that the mantle is stratified, with the upper and lower mantle convecting separately, and we estimate that the compositional density difference between these regions is about 5 (± 2) %. The stratification may not be perfect ('leaky layering'), but significant intermixing and homogenization of the upper and lower mantle over geological timescales are precluded.

Introduction

The purpose of this study is to interpret the density distribution through the Earth's lower mantle in terms of the bulk composition of this region. Specifically, we examine the degree to which an upper-mantle composition can satisfy the observed properties of the lower mantle. The reason for doing this is that whereas the uppermost mantle can be directly sampled and examined through the analysis of xenoliths and other rock fragments that are brought volcanically or tectonically to the surface, it is the lower mantle that makes up the bulk of our planet: almost 50% by mass and over 61% on an atomic basis. The problem, therefore, is to decide whether or not the outermost regions of the Earth that can be sampled are representative of the silicate portion of our planet, most of which cannot be directly observed.

We emphasize the analysis of density, rather than elastic moduli or wave velocities, because this is a relatively well-known property both for the Earth's interior and for candidate mineral assemblages existing at the pressures and temperatures of the deep mantle. The density distribution is mainly determined from the Earth's normal-mode frequencies and from the integral constraints of mean density and moment of inertia. We take the values of density in the Preliminary Reference Earth Model (PREM) (Dziewonski & Anderson 1981) and make comparisons over most of the depth range of the lower mantle. By sacrificing depth resolution in this way the density becomes accurately determined; averaged over a depth interval exceeding 1000 km, we expect the density to be known to better than 0.3–0.5% (Gilbert *et al.* 1973). In fact, the analysis rests mainly on the density distribution between depths of 1000 and 2000 km because the shallower regions may be somewhat heterogeneous or anomalous, as may be deeper regions (especially the D″ layer), and the equations of state of minerals become relatively uncertain at the pressures of depths exceeding 2000 km (see Dziewonski & Anderson 1981; Young & Lay 1987).

We begin by considering a model composition for the upper mantle, which is summarized in table 1 (all iron is listed as Fe^{2+}). This composition is derived mainly from analyses of mantle xenoliths, in addition to studies of ophiolite sequences and basalt petrogenesis (see, for example, Ringwood 1975; Yoder 1976; Green *et al.* 1979; Basaltic Volcanism Studies Project 1981). It corresponds to a garnet peridotite with roughly 55% olivine ($(Mg,Fe)_2SiO_4$) and 45% pyroxenes ($(Mg,Fe)SiO_3$ and $CaMgSi_2O_6$) and garnet (mainly $(Mg,Fe)_3Al_2Si_3O_{12}$), but because the olivine content is not critical for our analysis we allow the ratio of olivine to olivine and pyroxene to vary between $\frac{1}{2}$ and $\frac{2}{3}$. In contrast, the density is very sensitive to the iron content. In table 1, the ratio of MgO to MgO and FeO components is $x_{Mg} = 0.90\ (\pm 0.02)$. Although the majority of upper-mantle xenoliths yield $x_{Mg} = 0.90$–0.92 (Aoki 1984), we concentrate on the denser, iron-enriched limit ($x_{Mg} = 0.88$) for two reasons. First, the effect of partial melting is to deplete the iron out of the source region, so we are emphasizing the more fertile, unmelted compositions that represent the most primitive upper-mantle compositions that have been sampled. Second, the iron-rich composition comes closest to satisfying the observed density of the lower mantle. In fact, if the upper mantle actually has a composition with $x_{Mg} = 0.80$–0.86, our main conclusions either do not hold or cannot be proven with any certainty.

TABLE 1. MODELS OF UPPER-MANTLE COMPOSITION

oxide component	mass fraction (%)	element	atomic fraction (%)
SiO_2	45.0 (±1.4)	O	58.3
TiO_2	0.18 (±0.05)		
Al_2O_3	4.5 (±1.5)	Mg	20.1 (±1.0)
Cr_2O_3	0.4 (±0.1)	Si	15.8 (±0.5)
FeO	7.6 (±1.7)		
MnO	0.11 (±0.05)	Fe	2.2 (±0.5)
NiO	0.23 (±0.05)	Al	1.9 (±0.6)
MgO	38.4 (±2.0)	Ca	1.2 (±0.4)
CaO	3.3 (±1.0)		
Na_2O	0.4 (±0.2)	Na	0.27 (±0.14)
K_2O	0.01 (±0.1)	Cr	0.11 (±0.03)
total	100.1	Ni	0.06 (±0.01)
		Ti	0.05 (±0.01)
		Mn	0.03 (±0.01)
		K	0.004 (±0.04)

We note that the iron content is important mainly because of the large atomic mass of Fe: 55 against an average atomic mass of 21 for the composition in table 1. Thus, the density is controlled by the distribution of MgO, FeO and SiO_2 components, with the other constituents being present in small enough concentrations to be unimportant. The most abundant minor components, CaO and Al_2O_3, probably exist in a perovskite structured mineral at lower-mantle conditions, as do MgO, SiO_2 and (to a lesser degree) FeO. Whether the CaO and Al_2O_3 components enter into the Mg–Fe perovskite or form a separate perovskite (Liu & Bassett 1986), their contribution to the overall density appears to be negligible, according to currently available estimates (Jeanloz & Knittle 1986). Therefore, we evaluate the lower mantle composition in terms of the primary components MgO, FeO and SiO_2.

Mineralogy and temperature of the mantle

It is now well established that the olivine, pyroxene and garnet minerals existing in the upper mantle transform to denser phases when taken to the pressures of the transition zone and lower mantle (Jeanloz & Thompson 1983; Liu & Bassett 1986). The olivine is known to transform to the β-phase (spinelloid) and γ-spinel structures at the conditions of the transition zone (400–670 km depth), ultimately breaking down to a mixture of perovskite ($(Mg,Fe)SiO_3$) and magnesiowüstite ($(Mg,Fe)O$) at lower-mantle pressures. Over the same pressure interval, the pyroxenes and garnets react to form a silica-rich garnet (majorite) that trasforms to silicate perovskite at pressures existing near the top of the lower mantle. Thus, whether considered individually or as an assemblage, the dominant minerals of the upper mantle all form a silicate–perovskite assemblage at the conditions of the lower mantle.

Recent experiments demonstrate that $(Mg,Fe)SiO_3$ perovskite is stable throughout the entire pressure range of the lower mantle (Knittle & Jeanloz 1987). Therefore, we consider assemblages within the compositional range $2(Mg,Fe)SiO_3$ perovskite + $1(Mg,Fe)O$ magnesiowüstite (olivine:pyroxene ratio of 1) and $3(Mg,Fe)SiO_3$ perovskite + $2(Mg,Fe)O$ magnesiowüstite (olivine:pyroxene ratio of 2). The bulk composition is then completely specified by the olivine:pyroxene ratio and the x_{Mg} value, and the individual compositions of the perovskite and magnesiowüstite phases are obtained from the pressure-independent and temperature-independent partition coefficients of Bell *et al.* (1979) and Ito & Yamada (1982).

Because temperature and bulk composition together determine the density of a mineral assemblage, we consider the existing constraints on temperature in the mantle (figure 1). In the uppermost mantle, petrological studies offer the most direct estimates of temperature at depth (Jeanloz & Morris 1986). For example, the compositions of pyroxenes and garnets coexisting

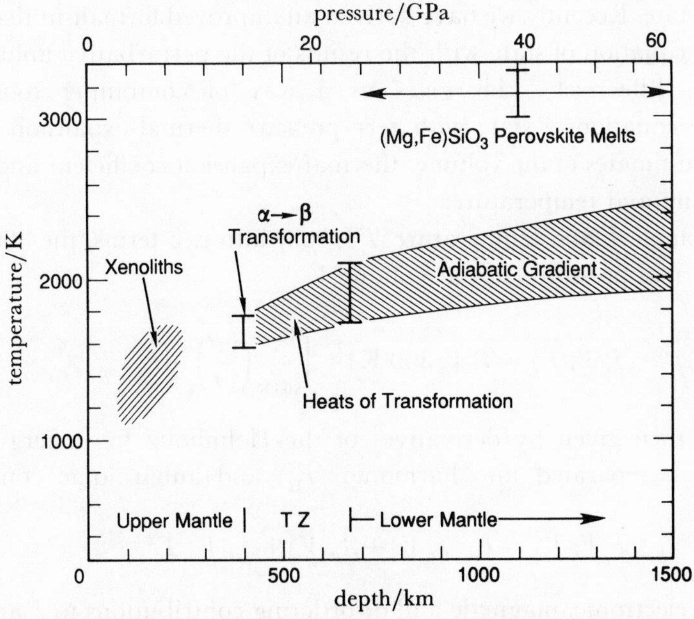

FIGURE 1. Constraints on temperature in the Earth's mantle shown as a function of depth (lower scale) and corresponding pressure (upper scale). Details are given in the text (see also Jeanloz & Morris 1986). TZ refers to the transition zone.

in xenoliths yield temperatures of 1600 (± 200) K for a depth of 200 km. The seismologically observed discontinuities at 400 and 670 km depth also provide estimates of temperature if these discontinuities can be ascribed to equilibrium phase transitions. Both the pressure at 400 km depth and the changes in velocities across it are compatible with the 400 km discontinuity being caused by the transformation of olivine to β-phase (Weidner & Ito 1987). Therefore, considering experimental determinations of the transition pressure as a function of temperature yields an estimate near 1700 K for the top of the transition zone at 400 km depth (Jeanloz & Thompson 1983; Navrotsky & Akaogi 1984; Bina & Wood 1987). Adding the heats of transformation for the breakdown to a perovskite assemblage finally leads to a temperature at 670 km depth of 1950 K (Jeanloz & Thompson 1983; Navrotsky & Akaogi 1984).

This estimate is based on the assumption that there is convection through the transition zone: the adiabatic temperature change with depth is included, but no consideration is given to the possible contribution from a thermal boundary layer at the top of the lower mantle (see Jeanloz & Richter 1979). With this assumption, the extreme limits on the temperature at 670 km depth are 1720–2100 K. Continued extrapolation along an adiabatic gradient therefore provides a minimum estimate of temperatures within the convecting lower mantle. At 1500 km depth a temperature of 2200 K, with an absolute minimum of 1920 K, is obtained in this way (figure 1). For comparison, the melting experiments of Heinz & Jeanloz (1987) place an upper bound of 3000–3300 K for depths of 700–1500 km in the mantle.

Thermal equation of state

The effects of temperature and pressure on the volumes (hence the densities) of both perovskite and magnesiowüstite have been measured separately. To combine the data, interpolate to the simultaneous effects of pressure and temperature, and extrapolate the results to the conditions of the deep mantle it is necessary to use a physically based theory for the thermal equation of state. Recently we have derived an improved formalism that combines the eulerian finite-strain equation of state with the results of the perturbative anharmonic theory of lattice dynamics. Effectively, this gives us a way of combining room-temperature measurements of the equation of state with zero-pressure thermal-expansion data to obtain internally consistent estimates of the volume, thermal-expansion coefficient and bulk modulus as functions of pressure and temperature.

The pressure at volume V and temperature T is split into two terms, the 300 K isothermal pressure and the thermal pressure (P_{th}) at volume V

$$P(V, T) = P(V, 300\text{ K}) + \int_{300\text{ K}}^{T} \left(\frac{\partial P}{\partial \tau}\right)_V \mathrm{d}\tau. \qquad (1)$$

The second term is then given by derivatives of the Helmholtz free energy, $(\partial P/\partial T)_V = -\partial^2 F/\partial V \partial T$, which is separated into harmonic (F_H) and anharmonic contributions (see Wallace 1972)

$$F(V, T) = F_H(T, V) + A_0(V) + A_2(V)\, T^2. \qquad (2)$$

For the present case, electronic, magnetic and disordering contributions to F are ignored, and the A_0 term is lost in the differentiation with respect to temperature. Thus, our analysis is limited to vibrational contributions to the thermal properties.

The philosophy behind this approach is to attempt only to fit data at temperatures of 300 K and above. Therefore, the dependency of F_H on the lattice vibrational spectrum is minimal and we consider only an average frequency, $\bar{\omega}$: more than a single characteristic frequency cannot be resolved in our analysis (see Jeanloz 1987). We identify $\bar{\omega}$ with the Einstein frequency (Einstein temperature $\theta_E = \hbar\bar{\omega}/k$), and set the corresponding Debye temperature to $\theta_D = \sqrt{\frac{5}{3}}\theta_E$ (Wallace 1972). With this correspondence between Einstein and Debye models we can verify that our data analysis is insensitive to the particular vibrational spectrum that is used, depending only on the characteristic frequency. We find this approximation to hold quite accurately (well within the quoted uncertainties) and can therefore set the thermal pressure to

$$P_{th}(T, V) = (\bar{\gamma}/V)\,[E_H(T, V) - E_H(300\text{ K}, V)] - \left(\frac{\partial A_2}{\partial V}\right)_T (T^2 - (300\text{ K})^2). \qquad (3)$$

Both Einstein and Debye models are used for the harmonic internal energy, E_H, and the average Grüneisen parameter is defined by

$$\bar{\gamma} = (-\partial \ln \bar{\omega}/\partial \ln V)_T. \qquad (4)$$

The advantage to this approach is that it combines two, physically based formulations that are empirically known to successfully describe existing measurements. These are the Eulerian finite-strain equation of state (see, for example, Birch 1952, 1978; Jeanloz & Knittle 1986; Jeanloz & Grover 1988; Jeanloz 1988, 1989) and the anharmonic description of thermal properties (Leibfried & Ludwig 1961; Zharkov & Kalinin 1971; Wallace 1972). The resulting expression for thermal expansion involves a more realistic equation of state (strain energy–volume relation) and, because the reference temperature is 300 K rather than 0 K, less sensitivity to the detailed vibrational spectrum than previous formulations, such as Suzuki's (1975). This is desirable for geophysical applications to high temperatures and pressures, but would not be so for low-temperature applications.

We fit (1) and (3) by first deriving the values of bulk modulus (K_0) and its pressure derivative (K_0') that reproduce the isothermal equation of state according to Birch's finite-strain formalism (truncated to third order in strain; see Birch 1978, for example). Then, the zero-pressure thermal expansion data are fitted to (1) by setting $-P(V, 300\text{ K}) = P_{th}(T, V)$ (i.e. $P(V, T) = 0$) to yield values of θ_{0D}, $\bar{\gamma}_0$, q and $q_2 A_{20}$. The Grüneisen parameter describes the volume dependence of the characteristic temperature according to (4), subscripts zero indicate ambient conditions ($P = 0$, $T = 300$ K), and

$$q = \mathrm{d}\ln\bar{\gamma}/\mathrm{d}\ln V, \qquad (5)$$

$$q_2 = \mathrm{d}\ln A_2/\mathrm{d}\ln V \qquad (6)$$

are assumed constant and of order unity. Note that $\bar{\omega}$, θ_D, $\bar{\gamma}$ and A_2 depend directly only on volume in this approach; indirectly they depend on temperature through the thermal expansion, but this is only significant for $\bar{\omega}$ or θ_D. The entire thermal equation of state, $P(V, T)$, is then given by the V_0, K_0, K_0', θ_{0D}, $\bar{\gamma}_0$, q, and the product $q_2 A_{20}$. These parameters are obtained by nonlinear least-squares fitting, in which equation-of-state and thermal-expansion measurements are weighted by the reciprocal of their pressure-dependent and temperature-dependent variances.

The resulting values for silicate perovskite and magnesiowüstite are listed in table 2. Here, x_{Fe} ($= 1 - x_{Mg}$) is the amount of iron component, but it should be noted that most of the measurements on which the values are based were obtained from Mg-rich samples. Primarily, we used the $T = 300$ K static-compression measurements of Knittle & Jeanloz (1987) on $Mg_{0.88}Fe_{0.12}SiO_3$ perovskite, along with Jackson & Niesler's (1982) ultrasonic data on MgO to 3.0 GPa (from which a 300 K compression curve is derived). The $P = 0$ thermal expansion measurements are those of Knittle *et al.* (1986) on perovskite of the same composition and of Suzuki (1975) on MgO. Uncertainties in the measurements have been propagated to the values given in table 2, and the effects of varying composition (Mg/Fe ratio) on the properties are discussed by Jackson *et al.* (1978), Jeanloz & Thompson (1983) and Williams *et al.* (1988).

Figures 2 and 3 summarize the data for perovskite and magnesiowüstite, along with the pressure–volume relations derived from the parameters in table 2. Because the pressure depends almost entirely on volume, it is evident that the 300 K isotherm is of primary importance for constraining the thermal equation of state. In comparison, the harmonic and

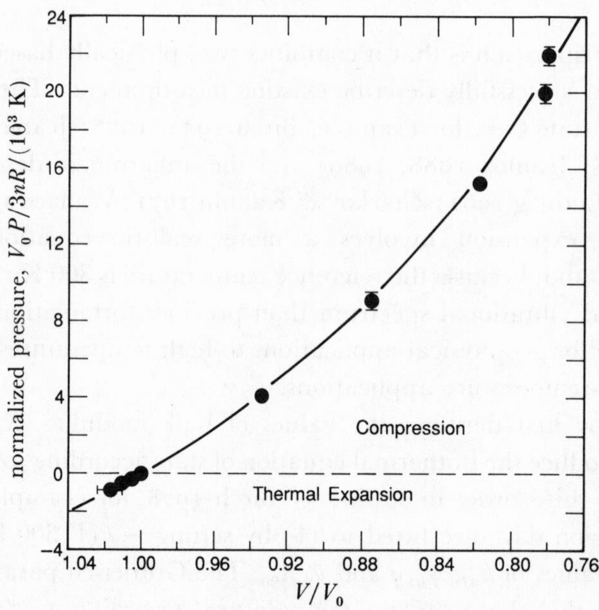

FIGURE 2. Thermal equation of state of $(Mg,Fe)SiO_3$ perovskite: summary of data (symbols with error bars) and model (curve) based on the parameters given in table 2 (see text).

TABLE 2. PHYSICAL PROPERTIES AT $P = 0$, $T = 300$ K

	perovskite $(Mg,Fe)SiO_3$	magnesiowüstite $(Mg,Fe)O$
volume, $V_0/(cm^3\ mol^{-1})$	$24.46\ (\pm 0.04) + 1.03\ (\pm 0.28) x_{Fe}$	$11.25\ (\pm 0.001) + 1.00\ (\pm 0.02) x_{Fe}$
atoms in formula, n	5	2
bulk modulus, K_0/GPa	266 (±6)	160.1 (±0.3)
pressure derivative, K_0'	3.9 (±0.4)	4.1 (±0.1)
Debye temperature, θ_0/K	725 (±25)	782 (±20)
Grüneisen parameter, γ_0	1.70 (±0.05)	1.66 (±0.02)
volume dependence, q	1 (±1)	1 (±1)
anharmonic term, $(q_2 A_{20}/3nR)/K^{-1}$	$-1.2\ (\pm 0.9) \times 10^{-4}$	$0.92\ (\pm 0.08) \times 10^{-4}$

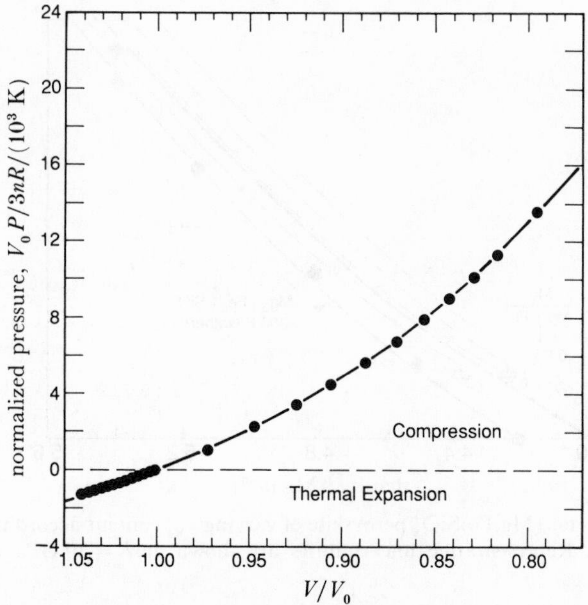

FIGURE 3. Thermal equation of state of MgO periclase: summary of data (symbols with error bars) and model (curve) based on the parameters given in table 2.

anharmonic thermal corrections are of secondary importance. We emphasize this point because it shows that to first order the density of the lower mantle is being compared with a measurement, not an extrapolation, of a mineral assemblage density at lower-mantle conditions. The importance of ultra-high-pressure measurements in this regard is also discussed by Bell *et al.* (1987).

RESULTS AND DISCUSSION

The thermal equation of state model for Mg–Fe silicate perovskite is illustrated in figure 4. Here the temperature is held constant at 2000 K, a plausible value for the deep mantle, and the effect of varying x_{Mg} is shown; the raw data making up the 300 K isotherm are also included for reference in this and the following two figures. Varying the iron content affects the density in two ways: primarily through the molar mass and secondarily through the molar volume (table 2). These effects are opposite in sign, but the first far outweighs the second.

The basic conclusion of our study is already evident in figure 4, even though this refers only to a pure pyroxene composition. For a typical upper-mantle value of $x_{Mg} = 0.90$, the density of perovskite is significantly less than that observed for the lower mantle. At 2000 K the discrepancy is 2 (± 1) %. As we point out below, this is a minimum value for the intrinsic density difference between the upper and lower mantle if these regions indeed differ in bulk composition. By intrinsic we mean a comparison at a given pressure and temperature. Although the effects on the density difference of varying temperature and silica content (ratio of olivine to pyroxene components) also need to be considered, these do not alter our initial conclusion.

Figure 5 summarizes the effect of temperature on the pressure–density relations of the four perovskite compositions that were considered in figure 4. It is clear that for a minimum estimated temperature in the deep mantle (*ca.* 1000–2000 km depth) of about 1900–2200 K,

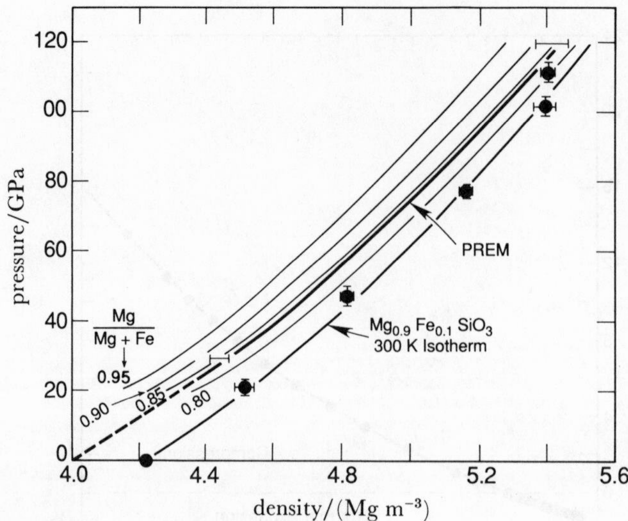

FIGURE 4. Isotherms (2000 K) for (Mg,Fe)SiO_3 perovskite of varying x_{Mg} content according to the present thermal equation of state model. Representative uncertainties are shown at $P = 30$ GPa and 120 GPa along the $x_{Mg} = 0.85$ isotherm.

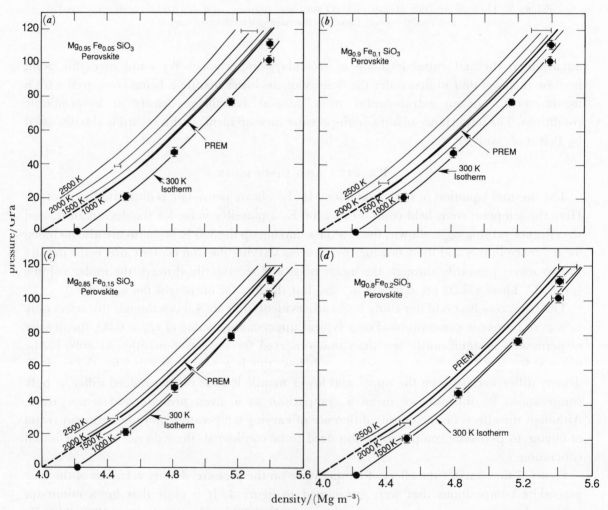

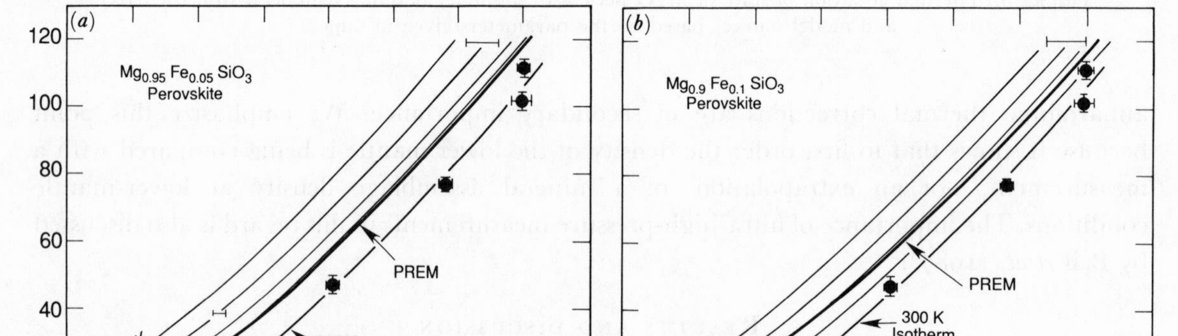

FIGURE 5. Isotherms for (Mg,Fe)SiO_3 perovskite according to the present thermal equation of state model, for compositions of $x_{Mg} = 0.95$ (*a*), 0.90 (*b*), 0.85 (*c*), and 0.80 (*d*). Representative uncertainties are shown at $P = 30$–40 GPa and 120 GPa along the $T = 2000$ K isotherm.

the largest acceptable value of x_{Mg} is in the range of 0.80–0.85. That is, considering only the silicate perovskite constituent and assuming no thermal boundary later at the 670 km discontinuity, a best estimate of $T = 2200$ K requires the ratio of MgO to MgO+FeO components to equal 0.79 (± 0.01). Although the lower-mantle geotherm is expected to be adiabatic, because this region is thought to be convecting vigorously, we only show isotherms for the pressure-density relations (see, for example, Jeanloz & Morris 1986). We do this both for simplicity and because the difference between an isotherm and an adiabat is probably less than our resolution. Nevertheless, there is a suggestion in figure 5 that higher temperature (hence greater iron content) leads to density profiles that are more compatible with the lower mantle being adiabatic. For example, the pressure–density relations make PREM appear to follow a sub-isothermal (let alone subadiabatic) trend in figure 5*a*, whereas it appears more nearly adiabatic in the comparisons shown in figure 5*c*, *d* (within the uncertainties).

The addition of magnesiowüstite (effectively, olivine component) hardly changes these calculations (figure 6). A limiting upper-mantle composition of $x_{Mg} = 0.88$ requires lower-mantle temperatures of about 1400 K, whereas a minimum estimate of 1900–2000 K for the lower mantle implies a composition of $x_{Mg} = 0.84$. These results are shown in figure 7, which summarizes from figures 5 and 6 the combinations of composition and temperature that are required to match the observed density of the lower mantle. Thus, a conventional estimate of upper-mantle composition ($x_{Mg} = 0.90$) would necessitate a temperature of 1200 (± 300) K in the lower mantle were the composition held constant.

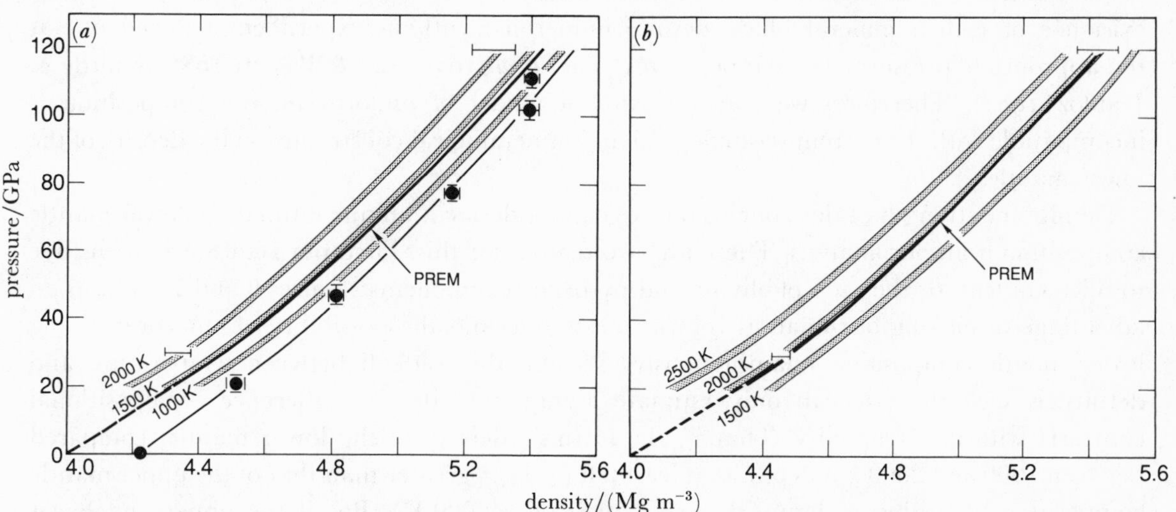

FIGURE 6. Isotherms for combinations of $(Mg,Fe)SiO_3$ perovskite and (Mg,Fe)O magnesiowüstite with total x_{Mg} of the assemblage being 0.88 (*a*) and 0.84 (*b*). Partition coefficients for Fe and Mg distributed between perovskite and magnesiowüstite are from Bell *et al.* (1979) and Ito & Yamada (1982). The width of shading for each isotherm reflects variations in composition between 2 perovskite+1 magnesiowüstite and 3 perovskite+2 magnesiowüstite.

Our attempt has been to be conservative in this analysis, tending to overstate the estimated uncertainties involved. Yet it is clear that if the upper mantle is more magnesium-rich than $x_{Mg} = 0.88$ and if the bulk of the lower mantle is at a temperature above 1900 K, then the lower mantle is intrinsically denser than would be expected for an upper-mantle composition: i.e. a model of uniform mantle composition. Unless the mineralogical constitution of the lower

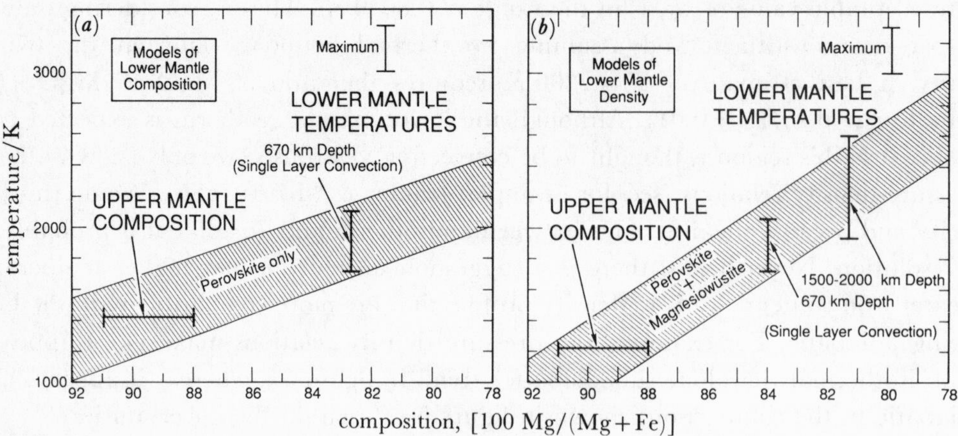

FIGURE 7. Tradeoff between composition (x_{Mg}) and temperature required to satisfy the observed density of the lower mantle. Assemblages of perovskite alone (*a*) and perovskite+magnesiowüstite in ratios between 2:1 and 3:2 (*b*) are shown.

mantle differs significantly from an assemblage dominated by $(Mg,Fe)SiO_3$ perovskite, as outlined above, the existing measurements of perovskite density at high pressures and temperatures lead inevitably to this conclusion. Specifically, one needs to invoke the presence in the lower mantle of an unknown, denser phase, that is either much denser than the perovskite or very abundant, to avoid this conclusion. So far, there is no evidence for the existence of such a mineral phase despite numerous synthesis experiments carried out at the appropriate pressures and temperatures (Yagi *et al.* 1979; Liu & Bassett 1986; Knittle & Jeanloz 1987). Therefore, we conclude that a model of uniform mantle composition is incompatible with the current seismological and mineralogical constraints on the density of the lower mantle.

Despite the strength of this conclusion, we cannot derive a unique estimate of lower mantle composition from our analysis. There are two reasons for this. First, our results are insensitive to silica content, or the ratio of olivine and pyroxene components (figures 6 and 7). This is an advantage in making our analysis robust, but it rules out the possibility of our constraining lower mantle composition unambiguously. Second, the tradeoff between temperature and density is such that we can only estimate a minimum density difference (compositional contrast) with any reliability. That is, the intrinsic density of the lower mantle, compared between 1000 and 2000 km depth, is at least 2.6 (± 1) % greater than that of the upper mantle composition according to figure 6*a* ($x_{Mg} \geqslant 0.88$, $T \geqslant 2000$ K). But if the upper and lower mantle differ in composition, they cannot intermix significantly over geological timescales, and they must therefore be separately convecting systems. In this instance of layered mantle convection, the temperature of the lower mantle would be expected to be at least 500 K higher than in the case of single-layer convection because of the presence of an extra thermal boundary layer near 700 km depth (Jeanloz & Richter 1979). The result is to increase the required contrast in intrinsic density between the upper and lower mantle to about 5 (± 2) %. In principle, the upper limit of 3000–3300 K for the temperature of the lower mantle (figure 1) bounds this value from above. In practice, however, the extrapolation of the equations of state becomes rapidly less certain with increasing temperature because the anharmonic contributions are not well determined at high pressures and temperatures. Therefore, it is difficult to uniquely identify the intrinsic density difference between the upper and lower mantle, and the compositional contrast is even less uniquely determined.

The arguments presented here are simplistic in that they ignore the lateral variations in composition, mineralogical constitution and temperature that must inevitably exist in the lower mantle (e.g. simply as a consequence of convective heat transfer). Therefore, we do not claim that the mantle must be rigidly layered. Locally dense or buoyant material can penetrate through the 670 km discontinuity if either the composition or temperature varies sufficiently to counteract the *ca.* 3–5 % intrinsic density contrast that we have estimated for the lower and upper mantle. In this sense the layering can be leaky, with the only limitation being that there has been insufficient mass transfer over geological history to completely homogenize the mantle. Because the original difference that may have existed between upper and lower mantle compositions is unknown, as is its evolution in time, there is no reason to rule out a finite (but incomplete) amount of intermixing between these regions.

Were there to be intermixing, the associated heat transfer would diminish the thermal boundary layer between the upper and lower mantle. Indeed, laboratory experiments suggest that partial, local intermixing may be inevitable at the interface between separately convecting fluids (H. Nataf, personal communication 1983; Olson 1984; Silver *et al.* 1988). In addition, there is some evidence that the viscosity of the lower mantle may be 1–2 orders of magnitude larger than that of the upper mantle (Hager 1984). Both the intermixing and the viscosity change are likely to smear out or at least obscure the presence of a thermal boundary layer. It is because of these complications that we have confined our analysis to a simple comparison of observed and modelled densities over the approximate depth range of 1000 to 2000 km, and assuming values of $x_{Mg} = 0.88$ for the upper mantle composition and 1900–2200 K for the temperature through most of the lower mantle.

Conclusions

Synthesis experiments at elevated pressures and temperatures indicate that the lower mantle consists almost entirely of $(Mg,Fe)SiO_3$ perovskite coexisting with (Mg,Fe)O magnesiowüstite. Thermal equations of state for these two minerals, either separately or in combination, are well constrained by existing measurements; in particular, the effect of pressure on density is well known up to the 100 GPa range. Consequently, the density distribution throughout the lower mantle can be modelled with little uncertainty. We find that if (i) the upper mantle is characterized by an iron/magnesium content of $x_{Mg} = 0.90$ (± 0.02), and (ii) the lower mantle is at a temperature of 2200 (± 300) K or above, then the observed density distribution through the lower mantle cannot be satisfied with an upper mantle composition (barring the presence of an unknown, sufficiently dense and abundant new phase). Therefore, the upper and lower mantle appear to differ in bulk composition. The intrinsic density difference between these regions is estimated to be about 3–5 %, with a minimum value of 2.6 (± 1) %, and this is sufficient to keep the mantle dynamically stratified (see Richter & McKenzie 1981; Olson 1984). These results suggest that the average temperature in the deep mantle exceeds 2200 K and that samples of the uppermost mantle are not representative of the bulk of the mantle.

We have benefited from discussions with T. G. Masters, D. L. Anderson, T. J. Ahrens and B. H. Hager, and from the comments of F. R. Boyd and D. Price. This work was funded by the

National Science Foundation. R.J. is grateful for the generous support provided by the Fairchild Scholar programme (California Institute of Technology) during the preparation of this manuscript.

References

Aoki, K. 1984 In *Materials science of the Earth's interior* (ed. I. Sunagawa), pp. 415–444. Tokyo: Terra Scientific.

Basaltic Volcanism Study Project 1981 *Basaltic volcanism on the terrestrial planets.* (1286 pages.) New York: Pergamon.

Bell, P. M., Mao, H. K. & Xu, J. A. 1987 In *High pressure research in mineral physics* (ed. M. Manghnani and Y. Syono), pp. 447–454. Washington: American Geophysical Union.

Bell, P. M., Yagi, T. & Mao, H. R. 1979 Iron-magnesium distribution coefficients between spinel $[(Mg,Fe)_2SiO_4]$, magnesiowüstite [(Mg,Fe)O] and perovskite $[(Mg,Fe)SiO_3]$. *Carnegie Inst. Wash. Yb.* **78**, 618–621.

Bina, C. R. & Wood, B. J. 1987 The olivine-spinel transition: experimental and thermodynamic constraints and implications for the nature of the 400 km seismic discontinuity. *J. geophys. Res.* **92**, 4853–4866.

Birch, F. 1952 Elasticity and constitution of the Earth's interior. *J. geophys. Res.* **57**, 272–286.

Birch, F. 1978 Finite strain isotherm and velocities for single-crystal and polycrystalline NaCl at high pressures and 300 °K. *J. geophys. Res.* **83**, 1257–1268.

Dziewonski, A. M. & Anderson, D. L. 1981 Preliminary reference Earth model. *Physics Earth planet. Inter.* **25**, 297–356.

Gilbert, F., Dziewonski, A. & Brune, J. 1973 An informative solution to a seismological inverse problem. *Proc. natn. Acad. Sci. U.S.A.* **70**, 1410–1413.

Green, D. H., Hibberson, W. O. & Jacques, A. L. 1979 In *The Earth: its origin, structure and evolution* (ed. M. W. McElhinny), pp. 265–299. New York: Academic Press.

Hager, B. H. 1984 Subducted slabs and the geoid: constraints on mantle rheology and flow. *J. geophys. Res.* **89**, 6003–6015.

Heinz, D. L. & Jeanloz, R. 1987 Measurement of the melting curve of $(Mg,Fe)SiO_3$ at lower mantle conditions and its geophysical implications. *J. geophys. Res.* **92**, 11437–11444.

Ito, E. & Yamada, H. 1982 Stability relations of silicate spinels, ilmenites and perovskites. In *High pressure research in geophysics* (ed. S. Akimoto & M. H. Manghnani), pp. 405–419. Tokyo: Center for Academic Publishing.

Jackson, I. & Niesler, H. 1982 The elasticity of periclase to 3 GPa and some geophysical implications. In *High pressure research in geophysics* (ed. S. Akimoto & M. H. Manghnani), pp. 93–113. Tokyo: Center for Academic Publishing.

Jackson, I., Liebermann, R. C. & Ringwood, A. E. 1978 The elastic properties of $(Mg_xFe_{1-x})O$ solid solutions. *Phys. Chem. Miner.* **3**, 11–31.

Jeanloz, R. 1987 Coexistence curves and equilibrium boundaries for high-pressure phase transformations. *J. geophys. Res.* **92**, 10352–10362.

Jeanloz, R. 1988 Universal equation of state. *Phys. Rev.* B. **38**, 805–807.

Jeanloz, R. 1989 Shock-wave equation of state and finite-strain theory. *J. geophys. Res.* (In the press.)

Jeanloz, R. & Grover, R. 1988 Birch-Murnaghan and U_s-up equations of state. In *Shock Waves in Condensed Matter* (ed. E. C. Schmidt & N. C. Holmes), pp. 69–72. New York: North-Holland.

Jeanloz, R. & Knittle, E. 1986 Reduction of mantle and core properties to a standard state by adiabatic decompression. *Adv. Phys. Geochem.* **6**, 275–309.

Jeanloz, R. & Morris, S. 1986 Temperature distribution in the crust and mantle. *A. Rev. Earth planet. Sci.* **14**., 377–415.

Jeanloz, R. & Richter, F. M. 1979 Convection, composition and the thermal state of the lower mantle. *J. geophys. Res.* **84**, 5497–5504.

Jeanloz, R. & Thompson, A. B. 1983 Phase transitions and mantle discontinuities. *Rev. Geophys. space Phys.* **21**, 51–74.

Knittle, E. & Jeanloz, R. 1987 Synthesis and equation of state of $(Mg,Fe)SiO_3$ perovskite to over 100 GPa. *Science, Wash.* **235**, 668–670.

Knittle, E., Jeanloz, R. & Smith, G. 1986 The thermal expansion of silicate perovskite and stratification of the Earth's mantle. *Nature, Lond.* **319**, 214–216.

Leibfried, G. & Ludwig, W. 1961 Theory of anharmonic effects in crystals. *Solid St. Phys.* **12**, 275–444.

Liu, L. & Bassett, W. A. 1986 *Elements, oxides and silicates.* (250 pages.) New York: Oxford University Press.

Navrotsky, A. & Akaogi, M. 1984 α-β-γ phase relations in Fe_2SiO_4–Mg_2SiO_4 and Ca_2SiO_4–Mg_2SiO_4: calculation from thermochemical data and geophysical applications. *J. geophys. Res.* **89**, 10135–10140.

Olson, P. 1984 An experimental approach to thermal convection in a two-layered mantle. *J. geophys. Res.* **89**, 11293–11301.

Richter, F. M. & McKenzie, D. P. 1981 On some consequences and possible causes of layered mantle convection. *J. geophys. Res.* **86**, 6133–6142.

Ringwood, A. E. 1975 *Composition and petrology of the Earth's mantle.* (618 pages.) New York: McGraw-Hill.
Silver, P. G., Carlson, R. W. & Olson, P. 1988 Deep slabs, geochemical heterogeneity and the large-scale structure of mantle convection: investigation of an enduring paradox. *A. Rev. Earth planet. Sci.* **16**, 477–541.
Suzuki, I. 1975 Thermal expansion of periclase and olivine, and their anharmonic properties. *J. Phys. Earth* **23**, 145–159.
Wallace, D. C. 1972 *Thermodynamics of crystals.* (484 pages.) New York: Wiley.
Weidner, D. J. & Ito, E. 1987 In *High pressure research in mineral physics* (ed. M. Manghnani & Y. Syono), pp. 439–446. Washington: American Geophysical Union.
Williams, Q., Knittle, E. & Jeanloz, R. 1988 Geophysical and crystal chemical significance of $(Mg,Fe)SiO_3$ perovskite. *Mineral Physics Monograph* no. 3. Washington D.C.: Am. Geophys. Union. (In the press.)
Yagi, T., Bell, P. M. & Mao, H. K. 1979 Phase relations in the system $MgO–FeO–SiO_2$ between 150 and 700 kbar at 1000 °C. *Carnegie Inst. Wash. Yb.* **78**, 614–618.
Yoder, H. S. Jr 1976 *Generation of basaltic magmas.* (265 pages.) Washington: National Academy of Sciences.
Young, C. J. & Lay, T. 1987 The core-mantle boundary. *A. Rev. Earth planet. Sci.* **15**, 35–46.
Zharkov, V. N. & Kalinin, V. A. 1971 *Equation of state for solids at high pressure and temperatures.* New York: Consultants Bureau.

Phil. Trans. R. Soc. Lond. A **328**, 391–407 (1989)
Printed in Great Britain

The properties and behaviour of mantle minerals: a computer-simulation approach

By G. D. Price[1], A. Wall[1] and S. C. Parker[2]

[1] *Department of Geological Sciences, University College London, Gower Street, London WC1E 6BT, U.K.*

[2] *Department of Chemistry, University of Bath, Bath BA2 7AY, U.K.*

The direct study of the majority of the physical properties of mantle-forming phases is currently beyond the limits of technology, because of the high pressures and temperatures (greater than 25 GPa and 1800 K) required to simulate lower-mantle conditions. As an alternative to direct study therefore, theoretical and computer-based techniques of lattice simulation and molecular dynamics have been employed to obtain an understanding of the behaviour of high-density silicates. The properties of perfect high-density silicate crystals, including their elastic and spectroscopic characteristics, have been investigated, but to date computer simulations of perfect lattice properties are insufficiently accurate to be used to solve geophysical problems. In contrast, the simulation of phase relations and defects properties are very successful. Small, negative Clapeyron slopes for perovskite-forming transformations are predicted, suggesting that the 670 km discontinuity may not be a rigid barrier to mantle convection. In addition, the activation energies for diffusion in forsterite and perovskite have been calculated, and the suggested high-temperature superionic conductivity of magnesium silicate perovskite has been confirmed.

Introduction

The aim of much recent research within the Earth sciences has been to determine the fundamental processes and mechanisms involved in mantle dynamics, lithospheric motion and plate tectonics. Many of the models, however, which have been advanced in an attempt to describe these vital processes rest upon unconfirmed assumptions concerning the response of rock-forming minerals to changes in their physical and chemical environment. This has led to the development of conflicting models for the behaviour of the Earth, epitomized by the two distinct and largely incompatible forms of model that currently exist to describe mantle convection; namely whole-mantle convection models and layered-convection models. The latter assume that a mineralogical phase transformation or chemical stratification occurring at depths close to 670 km below the crust, defines a boundary between the upper part of the mantle and a static or independently convecting lower mantle, whereas the former assume that no barrier to mantle convection exists. Strong arguments can be made in favour of both types of model, but it is neither possible to assess which best describes mantle convection, nor determine the effect of mineralogical or chemical changes on mantle behaviour, without knowing the properties of mantle-forming materials, such as magnesium silicate spinels, perovskites, garnets and ilmenites.

As a result of recent experimental research programmes, significant advances in our understanding of mantle-forming phases have been made, and although the nature of the

670 km discontinuity still remains unclear, the elastic constants (under ambient conditions) of mantle phases are now sufficiently well known that it is possible to conclude (figure 1) that the laterally averaged velocity and density structure of the lower part of the upper mantle and the transition zone (200–670 km) is at least consistent with, but does not demand, a chemically uniform, pyrolite mantle (see, for example, Weidner & Ito 1987; Irifune & Ringwood 1987). However, it is also widely recognized that the upper mantle, transition zone and lower mantle are significantly heterogeneous, with seismic velocity varying laterally by a few percent. The interpretation of these more subtle tomographic seismic observations provides a new challenge, as our current knowledge of the equations of state of mantle-forming phases is inadequate to establish unambiguously to what extent this heterogeneity is caused by thermal, chemical or crystal anisotropy effects. Similarly, our lack of understanding of the defect and rheological behaviour of these phases prevents us from developing a full description of mantle convection, the mantle viscosity profile, mantle conductivity (both thermal and electrical), the mantle geotherm, and the origins of the topography of the core–mantle boundary.

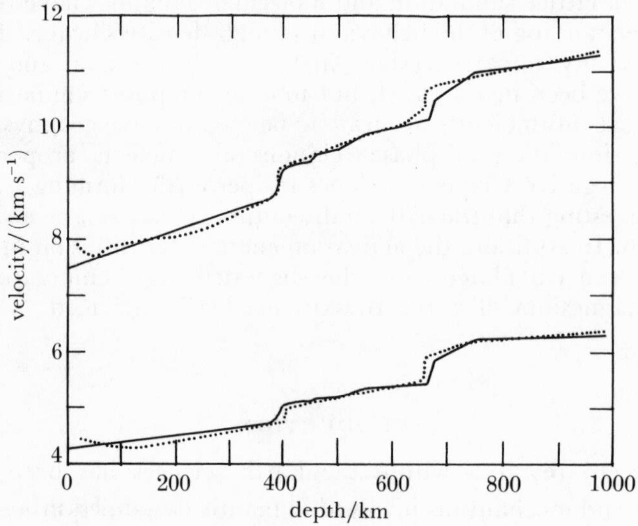

FIGURE 1. Calculated (solid lines) and observed (dotted lines) P- and S-wave velocities as a function of depth in the Earth's mantle, after the work of Weidner & Ito (1987).

Future progress in resolving these important geophysical problems depends upon being able to obtain highly accurate values for the seismic and defect properties of minerals, under pressure and temperature conditions appropriate to the mantle. To reproduce this environment experimentally represents a considerable challenge in itself, but one that has been met in recent years by the use of apparatus such as the diamond anvil cell (DAC) and the uniaxial split-sphere large-volume cell. However, to make *in situ* physical property measurements under these conditions, with the accuracy needed to be able to resolve existing geophysical problems, represents a challenge that is still to be overcome. Thus, the pressure and temperature derivatives of the elastic constants of mantle phases, their thermal-expansion coefficients, and their conductivity are as yet poorly constrained, and will be extremely difficult to measure with the accuracy that we require. It is the fact that there is little prospect of these experimental constraints being readily overcome that has led to the recent development of theoretical and computer-based, atomistic models of the behaviour and properties of high-density mantle-forming silicates (for recent reviews see Burnham 1985; Catti 1986; Catlow 1988). The aim of these computer-based studies is to provide an alternative way of establishing the properties of silicates at extremes of pressure and temperature, and to elucidate the microscopic or atomistic characteristics of mantle materials that determine their bulk, macroscopic or thermodynamic behaviour. In this paper, we will outline some of the approaches to the computer simulation of mantle-forming phases that have been adopted in recent years. We will attempt to highlight both the strengths and limitations of current techniques, and will discuss the course of possible future developments in the subject.

Approaches to silicate simulation

In principle, the bonding and related physical properties of any silicate can be established by the direct solution of the Schrödinger equation, which describes the interactions of electrons and nuclei within a given system. However, because of the complexity of most silicates, such quantum mechanical studies are usually limited to modelling simple structures or structural fragments. In contrast, the atomistic approach to modelling the behaviour of crystals is somewhat more simple and approximate, as it attempts only to describe the interactions between individual atoms or ions in the structure, rather than explicitly describing the interactions between each and every electron in the solid. This simplicity, however, allows the atomistic approach to be used to predict a wide range of physical and defect properties of crystals, while still providing useful insights into the nature of bonding within solids. In the atomistic approach, sets of interatomic potentials are developed, either *ab initio* or empirically, to describe the energy surface within a structure. The parametrization of the model potential is often designed to describe relatively simple, but effective concepts of chemical bonding, such as ionic interactions, short-range repulsion effects and van der Waals bonding.

The work of Cohen *et al.* (1988) exemplifies the current state of *ab initio*, total-energy calculations on crystalline silicates. These workers have used an approach based on the local-density approximation, derived from density functional theory, to perform total-energy calculations on cubic magnesium silicate perovskite. In their simulations they employ the linear augmented plane wave (LAPW) method to describe the charge density and potential surface within the unit cell. Although this approach is extremely accurate, it has the disadvantage that it is computationally very intensive, with the calculation of the energy of the cubic perovskite cell (containing only five atoms) taking between 10 and 20 central processing unit (CPU) hours on a Cray-XMP. Although useful in estimating the compressibility of the hypothetical cubic magnesium silicate perovskite, and revealing the predominantly ionic nature of the bonding in such highly coordinated silicates, this computationally intensive approach is not currently practical for the general study of geophysically important phases.

In contrast to performing total-energy calculations on silicate structures, Gibbs and co-workers (see Gibbs 1982; Lasaga & Gibbs 1987) have performed quantum mechanical simulations on 'silicate molecules', that can be envisaged as fragments of silicate minerals. Their calculations involve Hartree–Fock, *ab initio* solutions to the Schrödinger equation, which provide a potential energy surface that describes the interatomic interactions within the fragment. These calculations successfully model the known behaviour of true silicate molecules (such as disilic acid, $H_6Si_2O_7$), and the resulting potential energy surfaces have been parametrized and used to model simple silicates and silica polymorphs (Lasaga & Gibbs 1987).

A third approach, involving the potential induced breathing (PIB) model (Cohen *et al.* 1987;

Cohen 1987; Cohen *et al.* 1988), has also been successfully employed to model silicates. The PIB model avoids the need to solve the Schrödinger equation for the whole crystal, by employing the modified electron gas (MEG) model of Gordon & Kim (1972) to obtain interatomic potentials that can be used to simulate the overall potential energy surface within the crystal. In the MEG model, an approximate description of the crystal charge density is obtained by overlapping rigid ion charge densities, and the overall interatomic potential between two ions is obtained from their electrostatic interactions, their self-energy and the overlap energy between their charge densities. A problem exists, however, when modelling oxides with this approach, because of the fact that the O^{2-} ion is not stable in isolation. In MEG calculations, the O^{2-} ion must be stabilized by being contained within a charged sphere (the so-called Watson sphere), which simulates the stabilizing effect of the Madelung potential experienced by the oxygen ion when in the crystal lattice. In the PIB model, the oxygen interatomic potentials are parameterized in terms of the Madelung site potential; a change in the site potential induces a corresponding change in radius, or a 'breathing' of the Watson sphere. The PIB model potentials include not only pair-wise additive interactions but also many-body effects, and as a result, it successfully simulates trends in the Cauchy violations in alkaline earth oxides, that cannot be modelled by simple two-body potentials (Mehl *et al.* 1986).

Finally, the method that has been used most widely, and with considerable success, to describe the properties and behaviour of a large number of complex inorganic solids, including silicates, involves the use of empirical potential models (see Catlow 1987; Stoneham & Harding 1986). These potentials are empirical in the sense that they are described by an analytical form that is fitted to experimental data. The relative merits of empirical and *a priori* potentials depend on the proposed application. What is required in our geophysical studies is a quantitative prediction of physical properties, and as is pointed out by Stoneham & Harding (1986) 'at present the best empirical potentials have a far greater demonstrated accuracy than explicit solutions to the Schrödinger equation'. Indeed this point is echoed by Lasaga & Gibbs (1987), who point out the close agreement between their quantum mechanically derived Si–O potential and the empirical potential of Parker (1982), and conclude that the 'agreement supports the use of quantum mechanical calculations as a framework within which to shape interatomic potentials'.

In the empirical approach, sets of interatomic potentials are developed to describe the net forces acting upon atoms within a structure. The model potentials used can be polynomial fits to experimentally inferred potential energy surface, or can be based on functions designed to describe relatively simple concepts of chemical bonding. The ions in the structure can either be considered to be point charges, or can be viewed as having a more diffuse charge density. In either case, it appears that the electrostatic or Coulombic energy terms, which result from the ionic charges of the atomic species, are the most important component of the cohesive energy for ionic or semi-ionic solids such as silicates.

True ions are obviously not point charges, as is often assumed when calculating the Coulombic energy, but instead are composed of a nucleus and an associated electron cloud of finite size. In point-charge models, therefore, it is necessary to include a term in the potential that models the energetic effect of the overlap of the electron clouds, which results in a short-range repulsion effect that is most strongly felt by nearest-neighbour ions. Such short-range components of the two-body potential are well represented by the so-called Buckingham potential.

For fully ionic, rigid-ion models, the Coulombic and short-range terms are generally the only components of the potential to be considered. However, it is well established that bonding in a silicate, such as olivine, is not expected to be fully ionic, and in particular a degree of directional, covalent bonding between silicon and its coordinating oxygens is to be expected. In recent studies of silicates two approaches have been used in an attempt to model this more complex type of bonding. Price & Parker (1984) used non-integral or partial ionic charges, and used a Morse potential function to model the covalent Si–O bond. In contrast, Sanders *et al.* (1984) kept to full ionic charges, but modelled the directionality of the Si–O bond by introducing a bond-bending term into the potential. The use of a bond-bending term in the interatomic potential is a significant development, as it represents the consideration of effective three-body interactions within structure. Previous empirical models only considered purely two-body or pairwise additive effects, despite the fact that it has long been recognized that many-body interactions in crystal structures should be significant (Weidner & Price 1988). It is the success of potentials such as these that make possible the prospect of a detailed study of the equations of state of mantle phases.

It is often found that it is necessary to develop a so-called shell model (Dick & Overhauser 1958) as an alternative to the rigid-ion models described so far, to model correctly the dielectric properties of a crystal. A shell model provides a simple mechanical description of ionic polarizability, and it is therefore essential if the defect and high-frequency dielectric behaviour of a material are to be studied. In this model, the atom or ion (frequently it is assumed that oxygen is the only polarizable atom in the structure) is described as having a core containing all the mass, surrounded by a massless, charged shell, representing the outer valance electron cloud. The core and shell are coupled by a harmonic spring.

Computer modelling codes have been available for several years for the simulation of structural, elastic and dielectric properties of the perfect lattice and for calculations of defect energies, for a given set of interatomic potentials. Moreover, recent developments allow defect entropies and volumes to be calculated. The methods employ the 'effective potentials' discussed above, and provided that these are of adequate quality it is possible to perform reliable calculations of a wide range of perfect and defect lattice properties. In applying these methods to silicates, the main problem has been, and remains, obtaining reliable sets of potentials. The current status of the simulation of silicates will be reviewed briefly after we have summarized the principal types of methodology.

Simulation techniques

Full discussion of the techniques and the detailed theory behind them can be found in a number of standard papers (see, for example, Catlow & Mackrodt 1982; Catlow 1987), and so only a brief summary of the methods will be presented here.

Perfect-lattice simulations

In perfect-lattice simulations efficient summation techniques are used to evaluate the interatomic potentials, and to obtain the lattice energies of a compound and its first and second derivatives with respect to atomic coordinates. By coupling lattice-energy calculations with minimization routines, equilibrium crystal structures can be predicted. Moreover, these calculations can be performed to simulate the response of the structure to any required hydrostatic pressure. By using the first and second derivatives referred to above, it is possible to compute the predicted elastic, dielectric and piezoelectric constants of a structure, as well as all of its lattice vibrational properties (Catlow & Mackrodt 1982). From the lattice dynamical simulations, vibrational entropies and basic thermodynamic properties may be calculated by using standard relations (Born & Huang 1954), which when combined with the internal energy term, yield free energies. Successful applications of this technique have recently been reported by Price *et al.* (1987*a*, *b*) to the prediction of the thermodynamic properties and phase relations of magnesium silicate polymorphs.

Defect calculations

Defect energies may be calculated by using Mott–Littleton methods, the basis of which is the division of the crystal surrounding the defect into two regions: an inner region (containing typically 200–800 atoms) in which all atoms are relaxed to zero force; and an outer region whose response to the defect is treated by pseudo-continuum methods. There is now ample evidence available from work on halide and oxide crystals, that when suitable interatomic potentials are available these methods yield defect energies that agree very well with experimental values (see Catlow 1987; Catlow & Mackrodt 1982).

Defect formation leads to entropy changes due both to configurational and vibrational terms. The former is readily evaluated as $k \ln W$, where W is the orientational or site degeneracy associated with the defect. The latter is more difficult to determine; vibrational frequencies must be evaluated for both the perfect and defect lattice. However, recent developments allow this to be done by using both supercell and embedded crystallite methods. The calculations reported to date show the important role that entropy terms play in the total defect free energy.

Defect volumes are of vital importance in predicting the pressure-dependent defect-determined properties. The defect volume is defined by the thermodynamic relation

$$V = kV_0(\mathrm{d}G/\mathrm{d}V),$$

where k is the isothermal compressibility, V is the unit-cell volume and $(\mathrm{d}G/\mathrm{d}V)$ may be evaluated by performing calculations of the defect energies and entropies (and hence free energies) as a function of lattice parameter.

Atomic transport in solids is generally effected by defect migration processes. The rate of defect transport is governed by an Arrhenius expression of the type

$$A = A_0 \exp(-G^*/RT)$$

where G^* is the free energy of activation of the defect. Calculations of G^* are possible once the saddle-point for the migration process has been identified. In complex crystals this may involve an extensive search of the potential surface, a process that may require large amounts of computer time.

Simulating the effect of temperature

As discussed above, the simulation of temperature can be achieved by calculating the full lattice dynamical behaviour of a crystal. From this, standard statistical mechanics relations can be used to calculate heat capacity, entropy, thermal expansion coefficients, and other major thermodynamic properties of the material. The technique is limited, however, within the bounds of the quasiharmonic approximation, which leads to underestimates of high-temperature properties, because of the neglect of anharmonic effects. Such anharmonicity can only be studied via self-consistent phonon calculations or by molecular dynamics. Considerable progress has been made in developing a self-consistent lattice dynamics code (PARAPOCS), but further developments are required.

An alternative approach to the study of the effect of temperature is to use molecular dynamics simulation (MDS) techniques (see Dove 1988 for an extensive review). The approach was originally developed in the 1950s to study fluids, but it can also be used to model solids and has proved to be a useful technique for the study of superionic conductors and phase transformations. The advantage of using an MDS rather than a static stimulation is that it includes simulation of anharmonic temperature effects. Simulations of this type are, however, extremely costly in computer time.

The calculation is initiated with an array of atoms in a simulation box of fixed volume that must consist of a whole number of unit cells, but that may have any geometry. The forces on each ion are calculated according to the interatomic potential functions described above, and then Newton's equations of motion are solved to find their trajectory at all subsequent times by a series of iterations over a small time increment at a specified temperature.

Usually, because of the constraints on computer time it is only possible, to consider explicitly the trajectories of a few hundred particles. It is, therefore, necessary to eliminate surface effects. This is achieved by employing periodic boundary conditions in which the MDS box is embedded in an array of identical images extending to infinity. If a particle leaves the box through one face, its image enters with the same velocity through the opposite face. This does, however, impose an artificial periodicity on the system, and means that it is not possible for Schottky defects to form spontaneously during a simulation. Hence such defects must be introduced explicitly in setting up the simulation. Limitations on the amount of computing time available also normally dictate that simulations can only be run for 10–100 ps and, therefore, can only be used to study events that occur over this timescale. In addition, it is currently not feasible to include complex potentials, such as those that model polarizability.

Application and results

Equations of state

The early simulations of silicates concentrated on being able to reproduce the structures and elastic properties of known minerals. This requires being able correctly to describe both the first and second derivatives of the potential energy surface within the crystal. It was generally found, however, that either the structure could be reproduced at the expense of predicting elastic constants that were too stiff to be realistic, or that reasonable elastic constants could be obtained at the expense of predicting a structure with an unrealistically low density. However, with the realization of the importance of accounting for three-body or many-body interactions, it has become possible to be able to obtain models that predict zero-pressure, zero-temperature densities to within 2%, and elastic constants to within 5–10% of those determined experimentally. The fact that computer models can now be used accurately to predict the physical properties of mantle-forming phases was recently illustrated by the PIB model calculation of the elastic properties of magnesium silicate perovskite (Cohen 1987). These were published before experimental data were available (table 1), and have been subsequently shown to be in very close agreement with the values that were eventually determined, after considerable experimental problems had been overcome (Yeganeh-Haeri *et al.* 1988).

TABLE 1. OBSERVED AND PREDICTED STRUCTURAL AND ELASTIC PROPERTIES OF FORSTERITE AND MAGNESIUM SILICATE PEROVSKITE

(Cell parameters in ångströms (1 Å = 10^{-10} m = 10^{-1} nm), elastic constants in gigapascals. Observed data from Suzuki *et al.* (1983) and Yeganeh-Haeri *et al.* (1988). Calculated data at 300 K.)

	forsterite			perovskite			
	obs	THB1	THB2	obs	PIB	THB1	THB2
a	4.754	4.784	4.749	4.775	—	4.824	4.884
b	10.194	10.261	10.229	4.929	—	4.847	4.945
c	5.981	5.991	5.967	6.897	—	6.844	6.967
c11	328.6	357.1	320.0	520.0	531.0	660.9	600.8
c22	199.8	206.4	179.4	510.0	531.0	632.8	599.4
c33	235.5	280.3	248.9	437.0	425.0	551.2	485.5
c44	66.8	43.8	35.9	181.0	237.0	261.3	224.2
c55	80.9	74.0	73.2	202.0	249.0	251.1	220.6
c66	80.6	83.5	75.6	176.0	136.0	152.1	151.4
c12	66.7	92.1	66.4	114.0	44.0	147.4	100.2
c13	68.3	93.4	69.3	118.0	143.0	246.3	186.1
c23	72.6	87.8	62.7	139.0	166.0	245.3	176.1
K	128.9	154.5	128.1	245.0	249.0	347.0	290.0
G	81.4	78.2	73.3	184.0	192.0	213.0	200.7

TABLE 2. POTENTIAL FORM AND PARAMETERS USED IN THB1 AND THB2

(Units: A_{ij}, eV; B_{ij}, Å; C_{ij}, eV Å^6, $k_{\text{O-shell}}$, eV Å^{-2}, $k_{\text{O-Si-O}}$, eV rad^{-2}. Short-range cut off = 7.5 Å.)

$$U = \tfrac{1}{2}\Sigma[e^2 q_i q_j r_{ij}^{-1} + A_{ij}\exp(-r_{ij}/B_{ij}) - C_{ij} r_{ij}^{-6} - D_{ijk}(1+3\cos\theta_{ijk}\cos\theta_{jki}\cos\theta_{kji})(r_{ijk} r_{jki} r_{kji})^{-3} + k_{ijk}(\theta_{ijk}-\theta_0)^2 + k_i r_i^2]$$

	THB1	THB2		THB1	THB2
q_{Mg}	+2.0	+2.0	$q_{\text{O-shell}}$	−2.8480	−2.8480
q_{Si}	+4.0	+4.0	$q_{\text{O-core}}$	+0.8480	+0.8480
$A_{\text{Mg-O}}$	1428.5	875.0	$B_{\text{Mg-O}}$	0.2945	0.3225
$A_{\text{Si-O}}$	1283.9	1283.9	$B_{\text{Si-O}}$	0.3205	0.3205
$A_{\text{O-O}}$	22764.3	22764.3	$B_{\text{O-O}}$	0.1490	0.1490
$C_{\text{Si-O}}$	10.7	10.7	$C_{\text{O-O}}$	27.88	27.88
$D_{\text{O-Mg-O}}$	0.0	120.0	$D_{\text{O-O-O}}$	0.0	0.0
$k_{\text{O-shell}}$	74.9	74.9	$k_{\text{O-Si-O}}$	2.09	2.09

To illustrate the degree of accuracy that can now be expected from simulations, we also present in table 1 data for the structural and elastic properties of magnesium silicate perovskite and forsterite predicted by typical empirical or partially empirical potential models. Potentials THB1 and THB2 (table 2) are fully ionic models with short-range repulsive parameters empirically fitted to the structural and elastic properties of periclase (MgO) and quartz, and that are transferred to model the more complex silicate minerals considered here (see also Price *et al.* 1987*a*, *b*). Both potentials have polarizable oxygen ions. THB1 contains an O–Si–O three-body potential, whereas THB2 additionally contains an O–Mg–O three-body term. The importance of the O–Mg–O three-body term can be seen by its effect on the predicted elastic properties of forsterite. In the fully ionic description, its inclusion is necessary not only to obtain a reasonable prediction of the bulk modulus, but also to produce the correct prediction of the relative magnitudes of the shear and off-diagonal elastic constants (see also Weidner & Price 1988). Both potentials, however, are more successful in predicting the properties of forsterite than those of perovskite. This is probably because the coordination of Si and Mg are the same in forsterite as they are in the phases from which the potential parameters were derived, whereas in perovskite the coordination number of both ions is increased, with Si in octahedral coordination and Mg in 8–12-fold coordination. This problem illustrates one of the limitations of so-called transfer potentials, and is discussed more fully by Stoneham & Harding (1986). Other more accurate empirical, non-transferable potential models for magnesium silicate perovskite are, however, available (see Matsui *et al.* 1987; Wall 1988).

If potential models of silicates are to be geophysically useful, however, it is not sufficient for them to be able to reproduce the ambient properties of mantle minerals, they must also correctly predict their responses to changes in pressure and temperature. Cohen (1987) has shown that the PIB model prediction of the isothermal equation of state for magnesium silicate perovskite is in excellent agreement with experiment up to pressures in excess of 125 GPa (figure 2). PIB model predictions of the pressure derivative of the ambient temperature isentropic bulk modulus (K') and shear modulus (G') of 4.1 and 1.7, compare with the THB1 predicted values of 3.2 and 1.3, and the experimentally determined value of K' of 3.9 ± 0.4 (Knittle & Jeanloz 1987*a*). Similarly, the pressure derivatives of the bulk and shear moduli of forsterite predicted by THB2 are in accord with those determined experimentally, with predicted K' and G' values of 3.7 and 1.7 respectively, compared with accepted experimental values of 4.7 and 1.6 (Weidner 1986).

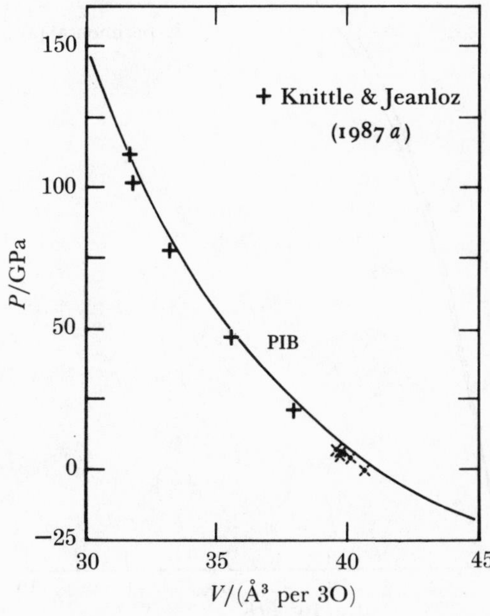

FIGURE 2. PIB model predicted and experimental equation of state for orthorhombic magnesium silicate perovskite at room temperature, after the work of Cohen (1987).

The predicted effect of temperature on the properties of silicates has recently been investigated by using quasiharmonic lattice dynamics (see, for example, Cohen 1987; Price *et al.* 1987*b*). From the calculation of the vibrational frequencies of the lattice, it is possible to obtain the heat capacity at constant volume (C_v) for the structure by using standard thermodynamic relations. The variation of the vibrational frequencies with pressure enables the mode Gruneisen parameters to be calculated which, in combination with predicted bulk modulus and heat capacity, enables the thermal expansion coefficient to be obtained. Price *et al.* (1987*a*, *b*) have shown that potential THB1 provides an excellent prediction of the lattice dynamical properties of forsterite, predicting infrared and Raman vibrational frequencies to within 30 wavenumbers and predicting phonon dispersion behaviour that is consistent with the limited experimental data that exists. The resulting predicted C_v curve (figure 3) is in outstanding agreement with experimental data; however, it should be noted that to obtain convergence the vibrational frequencies of at least eight points in the irreducible sector of the Brillouin zone must be used in the calculation. When the predicted thermal expansion coefficient (figure 4) is compared with measured values, however, it appears to be systematically underestimated by about 20%. This may reflect a shortcoming in the potential model used, or may reflect the fact that quasiharmonic calculations neglect anharmonic effects, which are known to be significant at high temperatures (see Ball 1986). This shortcoming could be overcome by developing a self-consistent phonons code or by using molecular dynamics to simulate these high temperatures.

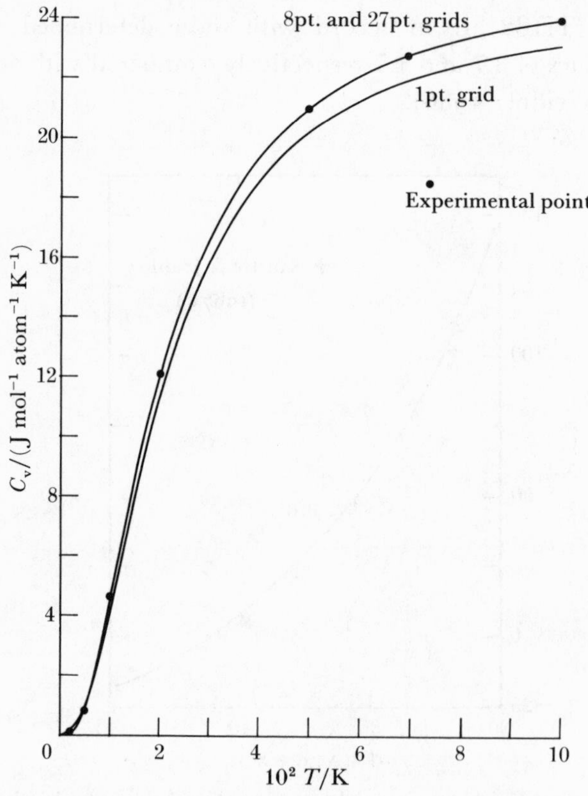

FIGURE 3. The THB1-predicted C_v curve for forsterite calculated with frequencies from 1, 8 and 27 point grids, from Price *et al.* (1987*b*).

Unfortunately, Cohen's PIB model potential for magnesium silicate perovskite is dynamically unstable, and imaginary frequencies are predicted. In addition, models such as PIB and the related MEG model used by Wolf & Bukowinski (1987) do not allow for anion polarizibility. As a result, it is impossible correctly to model LO–TO splitting in the infrared spectra of these silicates, and the predicted high-frequency lattice vibrations are seriously in error. As a result, the thermodynamic predictions of these models must be viewed with some caution.

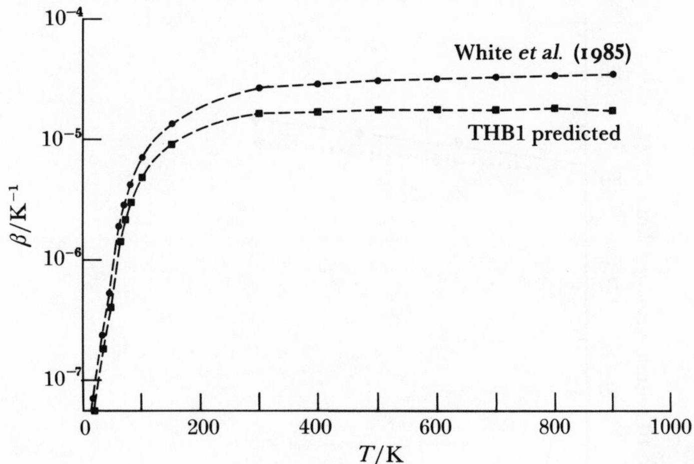

FIGURE 4. The observed and calculated thermal expansion coefficient of forsterite, from Price *et al.* (1987*b*).

Nevertheless, Cohen's model predicts a high-temperature (2000 K) coefficient of thermal expansion of 1.5×10^{-5} K^{-1}, and values of dK/dT and dG/dT of -0.014 and -0.007 GPa K^{-1} respectively, that are similar to those predicted by the dynamically stable THB1 and THB2 potentials. For magnesium silicate perovskite, these potentials predict a coefficient of thermal expansion of 1.6×10^{-5} K^{-1}, and dK/dT values of -0.027 and -0.009 GPa K^{-1} respectively. No experimental estimates of the temperature derivative of the elastic constants of perovskite exist; however, Knittle *et al.* (1986) have provided an estimate of the coefficient of thermal expansion of 4×10^{-5} K^{-1}. Recently, Matsui (1988) has published the results of a constant-pressure molecular dynamics simulation of perovskite, and predicted values of thermal expansion similar to those measured by Knittle *et al.* (1986). Because his study will have included all anharmonic effects, it is probable that Matsui's simulation provides a better model of the high-temperature behaviour of silicate perovskite than do those that are based on quasiharmonic simulations.

As discussed by Jackson (1983), Jeanloz & Thompson (1983), Weidner & Ito (1987) and others, if the chemical composition of the mantle is to be precisely defined, very accurate constraints must be placed upon the thermal expansion coefficients and elastic constants of mantle-forming phases. Weidner & Ito (1987) point out that a 10% uncertainty in K' for magnesium silicate perovskite, or an uncertainty of $\pm 2.0 \times 10^{-5}$ in its thermal expansion coefficient, has an equivalent effect on calculated seismic behaviour as changing the chemical composition of the lower mantle from pyrolite to chondritic. Thus, despite being able to predict seismic velocities with unprecedented accuracy (figure 5), we find ourselves in the frustrating position of still not being able to use these calculated equations of state to discriminate between various compositional models of the mantle.

Phase stability

In contrast to the limitations discussed above, computer simulations appear to be able to make an immediate contribution in the areas of the prediction of phases relations and of defect properties, where experimental results are poorly constrained. Thus, the experimental determination of the phase diagrams of the $(Mg,Fe)SiO_3$ and $(Mg,Fe)_2SiO_4$ systems are still

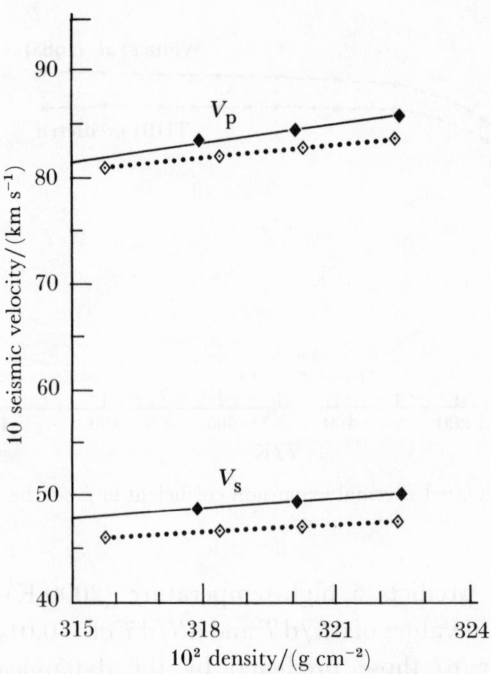

FIGURE 5. THB2-predicted and observed P- and S-wave velocities of forsterite as a function of density. Open symbols, calculated; filled symbols, from Suzuki *et al.* (1983).

subject to considerable uncertainty because of the difficulty in producing consistent pressure and temperature calibrations. The success that potentials such as THB1 have in reproducing the heat capacities of minerals, means that related thermodynamic properties, such as entropy, are also accurately predicted (Price *et al.* 1987*b*). By accurately simulating cell volumes and as well as free energies, these potentials can be used not only to predict the Clausius–Clapeyron slopes of phase boundaries (dS/dV), but also the whole phase diagram. Figure 6 (Marshall *et al.* 1988) shows the phase diagram for the Mg_2SiO_4 system as predicted by potential THB1 (coupled with an Si–O–Si bond bending term to simulate the Si_2O_7 group in the β-phase). The equations of the phase boundaries are very similar to those inferred experimentally (table 3), and confirm the shallow negative slope of the spinel/perovskite + MgO phase boundary. Comparable calculations on the $MgSiO_3$ system, predict the ilmenite/perovskite boundary to have a slope of -1.25 MPa K^{-1} in close agreement with experimentally derived values of between -1.2 and -2.5 MPa K^{-1}.

Following the fluid dynamical analysis of Christensen & Yuen (1984), such small negative Clapeyron slopes would be insufficient to cause these phase transformations to be a complete barrier to convection, unless the phase boundary also coincided with a chemical boundary that exhibited a chemical density contrast of greater than about 5%. Such a large chemical density contrast may be considered unlikely, but a slightly smaller chemical density contrast of *ca.* 2.5% would be compatible with current mineral physics data. In which case, the predicted Clapeyron slopes would suggest that significant penetration of a down going slab could be expected. Limited mixing between the upper and lower mantle could result therefore, but whole-mantle convection as such would not occur. Such a convective régime may indeed come closest to satisfying current mineral physics, seismological and geochemical constraints (Silver *et al.* 1987).

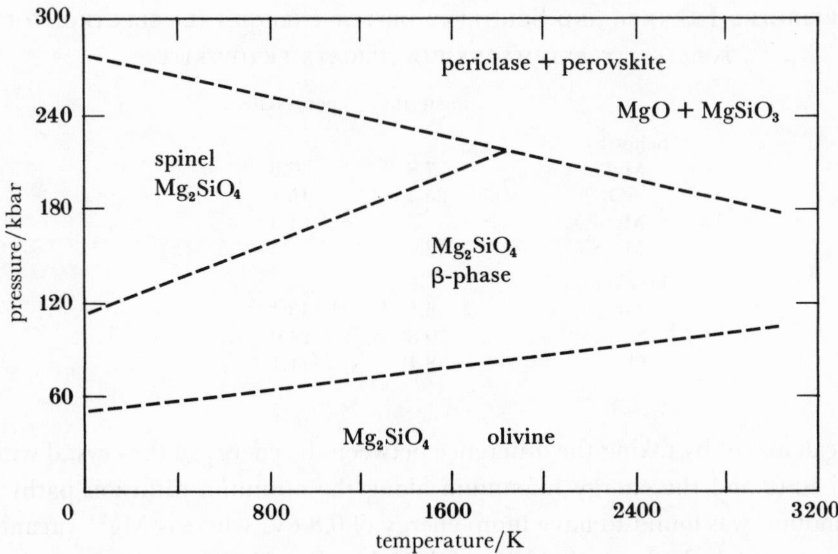

FIGURE 6. The calculated phase diagram for the magnesium orthosilicate system, from Marshall *et al.* (1988). (1 kbar = 10^8 Pa.)

TABLE 3. OBSERVED AND PREDICTED EQUATIONS FOR THE PHASE BOUNDARIES IN THE Mg_2SiO_4 SYSTEM, IN THE FORM $P = P_0 + aT$

(Units: P_0 in gigapascals, a in gigapascals per kelvin. Observed data from Ashida *et al.* (1987) and Ito & Yamada (1982).)

	observed	predicted
olivine – β-phase		
P_0	11.68	5.07
a	0.0025	0.0018
β-phase – spinel		
P_0	11.08	11.29
a	0.0048	0.0056
spinel – perovskite + MgO		
P_0	27.30	27.89
a	−0.002	−0.0033

Defect calculation

Because the viscosity and rheological behaviour of the mantle is likely to be determined by diffusion, controlled by intrinsic disorder within mantle-forming minerals, the calculation of the formation and migration energies of these intrinsic defects are of considerable importance. We have used potential THB1, in conjunction with the Mott–Littleton methodology discussed above, to model the defect behaviour of both forsterite and magnesium silicate perovskite (Wall & Price 1988). The predicted energies of Mg^{2+}, O^{2-} and Si^{4+} Frenkel defects pairs and a variety of Schottky defects are presented in table 4. In addition, we have examined the activation energies for magnesium vacancy and interstitial migration in forsterite, and for oxygen and magnesium vacancy migration in perovskite.

Calculations on forsterite indicate that Mg^{2+} Frenkel pairs are energetically most favourable, although a significant proportion of MgO Schottky defects would also be expected. The diffusion pathway for Mg^{2+} vacancies and interstitials was established by mapping out the potential energy surface within interstitial sites within the crystal. The jump energy for

Table 4. Predicted Frenkel and Schottky defect energies (in electronvolts) for forsterite and magnesium silicate perovskite

	forsterite	perovskite
Schottky		
MgO	7.8	7.6
SiO_2	25.4	15.1
$MgSiO_3$		21.3
Mg_2SiO_4	42.4	
Frenkel pair		
Mg	6.1	13.1
Si	30.8	24.0
O	8.4	11.1

diffusion was calculated by taking the difference between the energy of the crystal with a defect in the ground state and the energy maximum along the optimum diffusion pathway. Mg^{2+} interstitialcy motion was found to have jump energy of 0.8 eV, whereas Mg^{2+} vacancy motion has a jump energy of 1.1 eV. In the intrinsic régime therefore (i.e. at temperatures appropriate to mantle conditions), the predicted Arrhenius energy for Mg^{2+} diffusion is equal to the sum of half the Mg^{2+} Frenkel pair formation energy and the jump energy of the dominant diffusion process (in this case interstitialcy diffusion), and corresponds to a value of 3.85 eV. This value is in excellent agreement with experimental studies of Mg diffusion in olivine, which yield activation energies ranging between 3.7 and 4.5 eV (Anderson 1988). The success that we have in predicting the energetics of defects in forsterite leads us to believe that these potentials will also yield reasonable predictions of defect behaviour in perovskites.

In contrast to forsterite, Frenkel defects in the high-density perovskite structure are found to be energetically unfavourable, and the most common defect is predicted to be an MgO Schottky defect (table 4). With such a defect structure, it is probable that Mg and O diffusion in perovskite will occur by a vacancy migration mechanism. Assuming that O migration occurs by vacancy hopping, there are two possible pathways for O migration through the perovskite lattice: along the orthorhombic $\langle 100 \rangle_0$ direction (i.e. around the edges of the SiO_6 octahedra) or along the $\langle 110 \rangle_0$ direction (i.e. through the Mg_2O_4 octahedra). The jump energy for these two routes are predicted to be 0.8 eV and 3.2 eV respectively, and hence O migration along $\langle 100 \rangle_0$ is expected to be dominant. The Arrhenius energy of O diffusion in perovskite is expected, therefore, to be 4.6 eV in the intrinsic régime, but only 0.8 eV in the extrinsic régime. These calculations have been repeated at a simulated pressure of 100 GPa, and the activation volume for O diffusion has been calculated to be 1.6 $cm^3\ mol^{-1}$. The activation energy for Mg diffusion has also been calculated, for which it was assumed that the migrating Mg atom moved along a straight path through the centre of the Mg_2O_4 octahedron. The predicted jump energy for this process is 4.6 eV, and the resulting intrinsic Arrhenius energy is predicted to be 8.4 eV.

Perhaps not surprisingly, few experimental data exist on the energetics of diffusion processes in magnesium silicate perovskites. However, Knittle & Jeanloz (1987*b*) have reported the activation energy for the back-transformation of perovskite to pyroxene to be 70 ± 20 kJ mol^{-1} (0.7 ± 0.2 eV). Knittle & Jeanloz chose to interpret this energy in terms of the jump energy of Mg in the perovskite crystal. It is not obvious that this activation energy can be interpreted in terms of any jump process, but if it is, then we would suggest that it corresponds to extrinsic oxygen mobility (0.8 eV) rather than magnesium diffusion (4.6 eV). Knittle & Jeanloz

(1987*b*) also point out the importance of the activation volume in determining the effect of pressure on mantle diffusive processes. Our predicted value for the activation volume for O diffusion (1.6 $cm^3 \ mol^{-1}$) is sufficiently small to mean that oxygen is likely to be mobile even at the base of the lower mantle. Indeed, Yuen & Zhang (1987) recently calculated that to explain the inferred magnitude of the topography of the core–mantle boundary, it is necessary for the activation volume of the process that determines lower mantle viscosity to be lower than 2 $cm^3 \ mol^{-1}$. Our results suggest that oxygen migration would satisfy this criterion.

The movement of oxygen in the magnesium silicate perovskite lattice has also recently been studied by molecular dynamic simulations (Wall & Price 1988; Kapusta & Guillope 1988). The object of these studies has been to investigate whether magnesium silicate perovskite is a superionic conductor at mantle pressures and temperatures. According to some experiments carried out at atmospheric pressure, perovskites may exhibit anionic conductivities comparable to molten salts, and that the onset of superionic conductivity is associated with a significant reduction in viscosity (see Poirier *et al.* 1983). Should magnesium silicate perovskite also exhibit this behaviour, it would, therefore, have significant effects both on the electrical conductivity of the lower mantle and on its rheology. Although differing in detail, the general conclusion of both of these MDS studies is that at typical lower mantle temperatures and pressures (more than 2750 K and 100 GPa), where intrinsic defects would have a significant concentration, magnesium silicate perovskite is indeed predicted to be a superionic conductor, with a completely mobile oxygen sublattice. Wall & Price (1988) find that MDS supports their conclusion, drawn from static simulations, that oxygen diffusion occurs along $\langle 100 \rangle_0$. They also find that when the calculated diffusion coefficient of $0.4 \times 10^{-5} \ cm^2 \ s^{-1}$ is used in the Nernst–Einstein equation, it is predicted that the conductivity of magnesium silicate perovskite in the lower mantle will be approximately 100 $S \ m^{-1}$. This predicted conductivity is exactly that inferred for the lower mantle from the modelling of the temporal variations of the geomagnetic field (Ducruix *et al.* 1980), but is in apparent conflict with the experiments of Li & Jeanloz (1987), who failed to measure any electrical conductivity in a laser-heated, DAC-contained perovskite sample. Given the extreme experimental difficulty in making *in situ* electrical conductivity measurements, we must conclude that the electrical properties of magnesium silicate are still not fully resolved.

Conclusion

The progress that has been made in the development of computer simulations of silicates in the past five years has been remarkable. Nevertheless, it is not yet possible to use atomistic simulations to predict the thermoelastic properties of mantle-forming silicates with sufficient accuracy as to be able to use them to resolve the outstanding geophysical questions concerning the detailed composition of the mantle. To achieve this goal, it will be necessary to improve both the calculation and description of interatomic potential energy surfaces within minerals, and to improve upon the way in which anharmonic effects are treated in the calculation of high-temperature properties. Simulations of defects and kinetic processes are, however, already able to add to our understanding of mantle processes. In the future, it is to be hoped that it will be possible to investigate the role of Fe in determining behaviour of silicate perovskites, to investigate the onset of melting of mantle phases, to model silicate surfaces, and to extend the study of the effect of pressure on diffusion processes in silicates.

G.D.P. thanks the Royal Society for the receipt from 1983 to 1987 of a University 1983 Research Fellowship that enabled him to perform much of his work discussed in this paper. G.D.P. and A.W. also thank the NERC for support of their research programme. Maurice Leslie, Richard Catlow, Mark Doherty and Kate Wright are thanked for their help, advice and assistance.

References

Anderson, K. 1988 Ph.D. thesis, University of Clasthal.

Ashida, T., Kume, S. & Ito, E. 1987 In *High-pressure research in mineral physics* (ed. M. H. Manghnani & Y. Syono), Geophysical Monograph 39, pp. 269–274. Washington: American Geophysics Union.

Ball, R. D. 1986 *J. Phys. C.* **19**, 1293–1309.

Born, M. & Huang, K. 1954 *Dynamical theory of crystal lattices*. Oxford: Clarendon Press.

Burnham, C. W. 1985 In *Microscopic to macroscopic* (ed. S. W. Kieffer & A. Navrotsky), Reviews in Mineralogy 14, pp. 347–388. Washington: Mineralogical Society of America.

Catlow, C. R. A. 1987 In *Solid state chemistry: techniques* (ed. A. K. Cheetham & P. Day), pp. 231–278. Oxford: Clarendon Press.

Catlow, C. R. A. 1988 In *Physical properties and thermodynamic behaviour of minerals* (ed. E. K. H. Salje), NATO ASI Series C225, pp. 619–638. Dordrecht: Reidel.

Catlow, C. R. A. & Mackrodt, W. C. 1982 *Computer simulation of solids*, Lecture notes in physics, p. 166. New York: Springer-Verlag.

Catti, M. 1986 In *Chemistry and physics of terrestrial planets* (ed S. K. Saxena), Advances in Physical Geochemistry 6, pp. 224–250. New York: Springer-Verlag.

Christensen, U. R. & Yuen, D. 1984 *J. geophys. Res.* **89**, 4389–4402.

Cohen, R. E. 1987 *Geophys. Res. Lett.* **14**, 1053–1056.

Cohen, R. E., Boyer, L. L. & Mehl, M. J. 1987 *Phys. Rev.* **B35**, 5749–5760.

Cohen, R. E., Boyer, L. L., Mehl, M. J., Pickett, W. E. & Krakauer, H. 1988 In *Proceedings of the Chapman Conference on Perovskites*. Washington: American Geophysics Union. (In the press.)

Dick, B. G. & Overhauser, A. W. 1958 *Phys. Rev.* **112**, 90.

Dove, M. 1988 In *Physical properties and thermodynamic behaviour of minerals* (ed. E. K. H. Salje), NATO ASI Series C225, pp. 501–590. Dordrecht: Reidel.

Ducruix, J., Courtillot, V. & Mouel, J. L. 1980 *Geophys. Jl R. astr. Soc.* **61**, 73–94.

Gibbs, G. V. 1982 *Am. Miner.* **67**, 421–450.

Gordon, R. G. & Kim, Y. S. 1972 *J. chem. Phys.* **56**, 3122–3133.

Irifune, T. & Ringwood, A. E. 1987 In *High-pressure research in mineral physics* (ed. M. H. Manghnani & Y. Syono), Geophysical Monograph 39, pp. 231–242. Washington: American Geophysics Union.

Ito, E. & Yamada, H. 1982 In *High-pressure research in geophysics* (ed. S. Akimoto & M. H. Manghnani), Advances in Earth and Planetary Sciences, no. 12, pp. 405–420. Dordrecht: Reidel.

Jackson, I. 1983 *Earth planet. Sci. Lett.* **62**, 91–103.

Jeanloz, R. & Thompson, A. B. 1983 *Rev. Geophys. Space Phys.* **21**, 51–74.

Kapusta, B. & Guillope, M. 1988 *Phil. Mag.* (In the press.)

Knittle, E. & Jeanloz, R. 1987*a* *Science, Wash.* **235**, 669–670.

Knittle, E. & Jeanloz, R. 1987*b* In *High-pressure research in mineral physics* (ed. M. H. Manghnani & Y. Syono), Geophysical Monograph 39, pp. 243–250. Washington: American Geophysics Union.

Knittle, E., Jeanloz, R. & Smith, G. L. 1986 *Nature, Lond.* **319**, 214–216.

Lasaga, A. C. & Gibbs, G. V. 1987 *Phys. Chem. Mineral.* **14**, 107–117.

Li, X. & Jeanloz, R. 1987 *Geophys. Res. Lett.* **14**, 1075–1078.

Marshall, I., Parker, S. C. & Price, G. D. 1988 *J. chem. Soc. chem. Commun.* (In the press.)

Matsui, M. 1988 *Phys. Chem. Mineral.* (In the press).

Matsui, M., Akaogi, M. & Matsumoto, T. 1987 *Phys. Chem. Mineral.* **14**, 101–106.

Mehl, M. J., Hemley, R. J. & Boyer, L. L. 1986 *Phys. Rev.* **B33**, 8685–8696.

Parker, S. C. 1982 *UKAEA Report*, AERE TP963.

Poirier, J. P., Peyronneau, J., Gesland, J. Y. & Brebec, G. 1983 *Physics Earth planet. Inter.* **32**, 273–287.

Price, G. D. & Parker, S. C. 1984 *Phys. Chem. Mineral.* **10**, 209–216.

Price, G. D., Parker, S. C. & Leslie, M. 1987*a* *Mineralog. Mag.* **51**, 157–170.

Price, G. D., Parker, S. C. & Leslie, M. 1987*b* *Phys. Chem. Mineral.* **15**, 181–190.

Sanders, M. J., Leslie, M. & Catlow, C. R. A. 1984 *J. chem. Soc. chem. Commun.*, pp. 1271–1273.

Silver, P. G., Carlson, R. W. & Olson, P. 1987 *Eos, Wash.* **68**, 1500.

Stoneham, A. M. & Harding, J. H. 1986 *A. Rev. phys. Chem.* **37**, 53–80.

Suzuki, I., Anderson, O. L. & Sumino, Y. 1983 *Phys. Chem. Mineral.* **10**, 38–46.

Wall, A. 1988 Ph.D. thesis, University of London.
Wall, A. & Price, G. D. 1988 In *Proceedings of the Chapman Conference on Perovskite* (ed. A. Navrotsky & D. J. Weidner). Washington D.C.: American Geophysical Union. (In the press.)
Weidner, D. J. 1986 In *Chemistry and physics of terrestrial planets* (ed. S. K. Saxena), Advances in Physical Geochemistry no. 6, pp. 251–274. New York: Springer-Verlag.
Weidner, D. J. & Ito, E. 1987 In *High-pressure research in mineral physics* (ed. M. H. Manghnani & Y. Syono), Geophysical Monograph 39, pp. 439–446. Washington: American Geophysics Union.
Weidner, D. J. & Price, G. D. 1988 *Phys. Chem. Mineral.* (In the press.)
White, G. K., Roberts, R. B. & Collins, J. G. 1985 *High Temp. high Pressure* **17**, 61–66.
Wolf, G. H. & Bukowinski, M. S. T. 1987 In *High-pressure research in mineral Physics* (ed. M. H. Manghnani & Y. Syono), Geophysical Monograph 39, pp. 313–334. Washington: Amer. Geophys. Union.
Yeganeh-Haeri, A., Weidner, D. J. & Ito, E. 1988 In *Proceedings of the Chapman Conference on Perovskite* (ed. A. Navrotsky & D. J. Weidner). Washington: American Geophysics Union. (In the press.)
Yuen, D. & Zhang, S. 1987 *Eos, Wash.* **68**, 1488.

Phil. Trans. R. Soc. Lond. A **328**, 409–416 (1989)
Printed in Great Britain

Characteristics of perovskite-related materials

By E. Salje

Department of Earth Sciences, University of Cambridge, Downing Street, Cambridge CB2 3EQ, U.K.

Oxides with perovskite structure show characteristic features of their physical behaviour that can be relevant for the understanding of the mantle material $(MgFe)SiO_3$. Perovskites distort locally, and four different mechanisms have been found: tilt of octahedral complexes, off-centring of the octahedrally coordinated cation, distortion of the octahedral cage and off-centring of the 12-fold coordinated site. These deformations usually lead to structural instabilities related to phase transitions, extremely sluggish kinetic behaviour even leading to pseudo-glasses and polaronic transport. All these properties depend sensitively on the oxygen fugacities, defect densities and the rheology of the system. There is some evidence that the predicted structural phase transitions and glass states occur in $CaSiO_3$ whereas no structural instability has yet been reported in $MgSiO_3$.

Introduction

Materials with perovskite structure, or topologically closely related structures, have been of central interest to many solid-state scientists over the past three decades. Typical examples for research in physics range from the investigation of critical phenomena during structural phase transitions (e.g. in $SrTiO_3$ and $PrAlO_3$) to superconductivity in Cu-containing phases. Chemists are mainly interested in transport properties and defect structures (e.g. in WO_3 and related bronzes) and engineers and materials scientists have focused their research on electronic switching and the characterization of the properties of ceramic material. Research on perovskite structures has only recently been extended to $(MgFe)SiO_3$, a material that is thought to make up most of the lower Earth mantle. The relatively new research efforts on $(MgFe)SiO_3$ can potentially profit from the knowledge of the physical and chemical behaviour of other perovskite structures. It is the purpose of this paper to outline some of the typical materials properties of well-known oxides with perovskite structure.

Structural instabilities

The most outstanding feature of the perovskite structure is that its topology (i.e. the principal arrangement of atoms) is thermodynamically extremely stable whereas its actual crystal structure (i.e. the actual positions of the atoms) appears to be very unstable (Salje 1976*a*, *b*). This structural instability leads almost invariably to structural deformations that destroy the high symmetric atomic arrangements that would occur in cubic material with the space group *Pm*3*m*, (figure 1). In fact, virtually all perovskite structures are low symmetric, having lowered their symmetry via structural phase transitions. There are four typical deformation patterns, as shown in figure 1. They consist of octahedral tilts, off-centring of the octahedra midpoint B-position, octahedral distortion and shift of the 12-fold coordinated A-position away from the

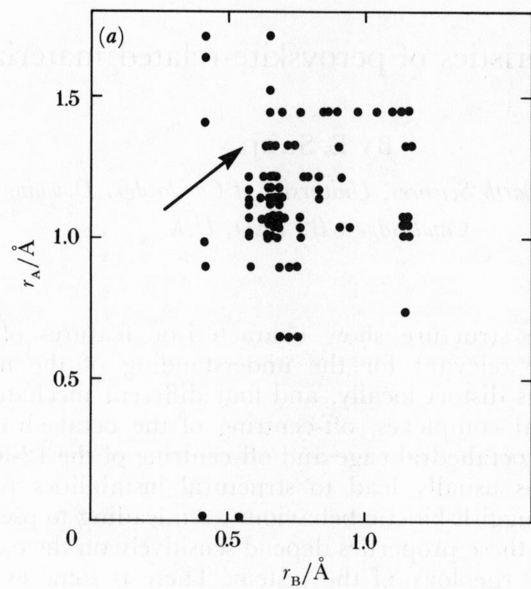

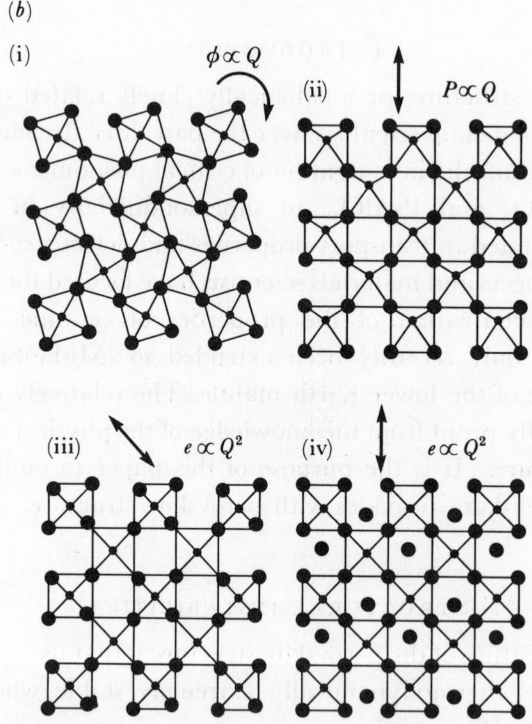

FIGURE 1. (*a*) Distribution ionic radii in various perovskites (r_A = radius of A atom, r_B = radius of B atom). The densest packing would occur for $r_A \approx 1.3$ Å, and $r_B \approx 0.56$ Å as indicated by the arrow. (1 Å = 10^{-10} m = 10^{-1} nm.) (*b*) Distortion pattern of perovskite structures: (i) tilt distortion (octahedra rotation): (ii) Slater distortion (off-centring of the B position; (iii) 'co-valent' distortion (distortion of the oxygen octahedra); (iv) off-centring of the A position. The order parameter Q is indicated in each case, note that (ii) is drawn as Γ-instability whereas all other examples are zone-boundary instabilities.

cavity centre. The crystal chemical origin for these structural distortions has been explored in great detail, both theoretically and experimentally (see, for example, Luthi & Rehwald 1981; Blinc & Zeks 1979; Muller 1986).

Changes in the external thermodynamic conditions, such as temperature, pressure, or chemical composition, lead to systematic structural variations that appear as collective movement of the atomic positions. This leads, in turn, to critical phenomena related to structural phase transitions. Such critical behaviour appears as strongly temperature dependent elastic constants, as in the case of $PrAlO_3$ (figure 2). Soft optic modes are a very common feature in tilt instabilities and phase transitions involving off-centring mechanisms (figure 3). The correlated order parameter follows mainly Landau-behaviour over a large temperature interval under non-critical conditions. Under temperature and pressure conditions close to the critical point (in the Ginzburg interval), fluctuation behaviour occurs for some perovskites (figure 4). The Ginzburg interval is, however, small enough to be ignored for most geological applications. It is important to note that all other physical parameters, such as density, spontaneous strain, specific heat, ultrasonic attenuation, domain formation and velocity of the domain walls, conductivity, ionic transport and cation ordering, are heavily influenced by the variations of this order parameter and do vary significantly, therefore, as a function of temperature and pressure (see reviews in Salje 1988*a*).

This surprising correlation between extreme topological stability and structural instability is superceded in some perovskite structures by their apparent unwillingness to complete a structural phase transition. This effect is essential for the application of many perovskite

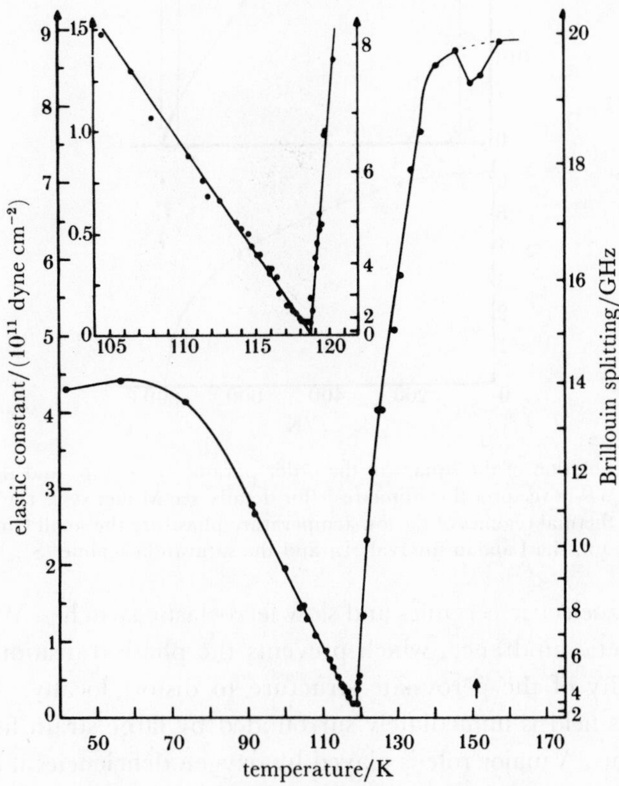

FIGURE 2. Temperature evolution of the elastic shear constant near the 118 K transition in $PrAlO_3$ (after Fleury *et al.* 1974).

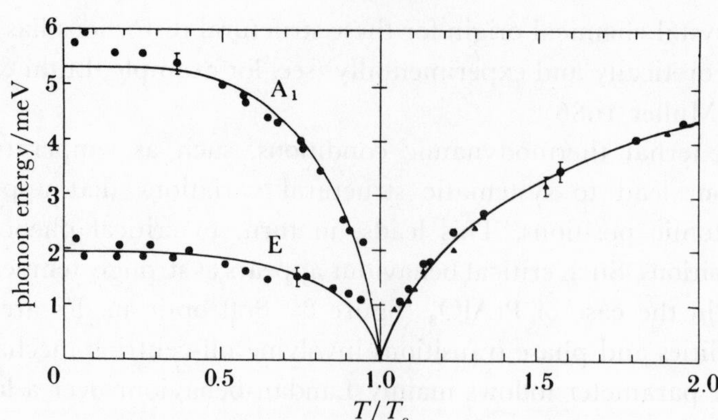

FIGURE 3. Soft-mode behaviour of $SrTiO_3$. The triply degenerate mode at $T > T_c$ splits into an A_1 mode and a double-degenerate E mode at $T < T_c$. (After Fleury *et al.* 1968; Cowely *et al.* 1969; Shirane & Yamada 1969.)

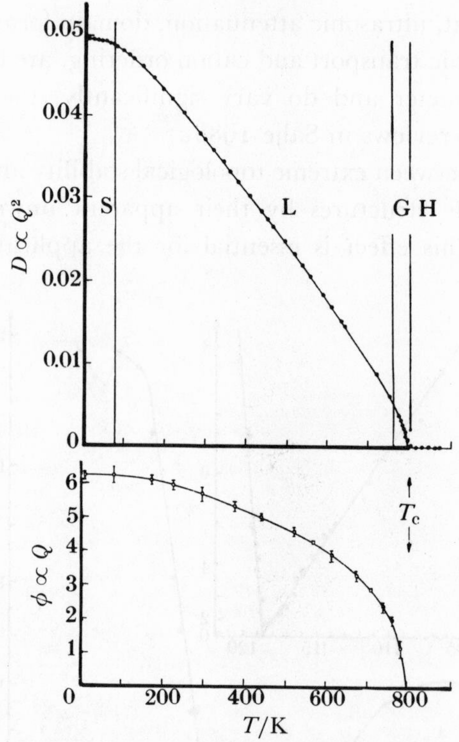

FIGURE 4. Temperature evolution of the square of the order parameter Q (top) and the linear order parameter (bottom) for $LaAlO_3$ (D in reciprocal centimetres) (for details, see Muller *et al.* 1968). The high-temperature phase is H, the three thermal régimes of the low-temperature phase are the small Ginzburg interval (G) where fluctuations are relevant, the Landau interval (L) and the saturation régime (S).

materials such as piezoelectric ceramics and slow ferroelastic switches. We can understand the sluggishness and kinetic hindrance, which prevents the phase transition occurring, from the extraordinary capacity of the perovskite structure to distort locally. Any impurity, lattice imperfection or stress field is immediately surrounded by large strain fields, which make the crystal inhomogeneous. A major role is played by oxygen deficiencies if a perovskite is placed in a slightly reducing environment. Its physical properties can thereby change dramatically. The term 'dirty ferroelectrics' has been introduced since as early as 1973 by Burns & Scott

(1973*a*) after the pioneering work of Russian groups around Smolensky and Levanyuk had penetrated western literature. The idea behind inhomogeneous, 'dirty' perovskites is that lattice imperfections couple with the order parameter of the structural phase transition. Those components of the distortion field that have the same symmetry as the order parameter act as its conjugated field and prevent the phase transition from occurring at the critical temperature. If the coupling is weak, the transition is smeared out over a large temperature interval. If the coupling is strong, the conjugate field leads to the decomposition of the crystal into interacting domains that can suppress the transition process altogether.

The most striking experimental observation for this behaviour is related to ceramic material or powdered samples. Typical examples are the commonly used $PbTiO_3$ ceramics which show no transition although the crystalline material does (Burns & Scott 1973*b*). Even more extreme is the behaviour of WO_3, which undergoes a major phase transition PI–P_c. This phase transition is completely suppressed if the material is pulverized in an agate mortar (Salje 1976*a*, *b*; Salje *et al.* 1978).

Kinetic behaviour

The tendency of the perovskite structure to distort on virtually any microscopic and submicroscopic lengthscale also leads to dramatic effects in their kinetic behaviour. Metastable structural states can often arise because the energy surface in the configuration space is highly structured. This is the outcome of many competing deformation schemes with virtually the same energy. The time dependence of a system travelling on this surface is simply defined by a generalized flow equation (Salje 1988*b*)

$$\frac{\mathrm{d}Q}{\mathrm{d}t} = -\frac{\gamma\langle a^2\rangle}{2k_B T}\left(1-\frac{\zeta_c^2}{\zeta^2}\,\mathrm{e}^{\xi^2/2\nabla^2}\right)\frac{\delta G}{\delta Q},$$

where Q is the macroscopic order parameter, and t is the time. The other essential parameter is the ratio ζ_c/ζ, which defines the distance (in units of the correlation length) over which the order parameter Q is conserved. In a simple phase transition, Q is a non-conserved quantity and changes to zero in the high symmetric phase. In spinodal decomposition, the order parameter is conserved as averaged over several modulations, as in the case of modulated phases. Many experimental results and some theoretical concepts have been worked out for the two extreme cases $\zeta_c/\zeta = 0$ and $\zeta_c/\zeta = 1$ (see Salje 1988*b*) but very little is known about intermediate cases that seem to be relevant for the understanding of perovskite structures under non-equilibrium conditions.

The non-equilibrium behaviour can be characterized by a simple example. A perovskite structure under inhomogeneous stress and large local temperature variations (as typical for most experiments in diamond anvil cells) will develop a transformation pattern that follows a simple kinetic law (Salje & Wruck 1988)

$$\langle Q\rangle = L(P) = P(x)\,\mathrm{e}^{-xt}\,\mathrm{d}x,$$

where L means the Lorentz transformation, P is the probability function and t is the time. A Kirkpatrick–Sherrington glass (Kirkpatrick & Sherrington 1978) follows

$$P(x) = \frac{1}{2\pi}\left(\frac{4}{x}-1\right)^{\frac{1}{2}},$$

whereas a pseudo-spin glass can be approximated by

$$P(x) = \frac{1}{x(x-2)}.$$

Both probability functions lead to inhomogeneous systems that will, only after very long times, equilibrate. Thermodynamic equilibrium might not be attainable under laboratory conditions. Natural material, on the other hand, could equilibrate and it is uncertain how far laboratory experiments under largely inhomogeneous temperature and pressure conditions can be paralleled with the large-scale behaviour of the lower mantle.

Electronic transport properties

The tendency of the perovskite structure to distort locally also leads to the formation of polaronic states, because charge carriers interact sufficiently strongly with the surrounding lattice that their mobility is reduced and self-trapping is obtained. The relaxation time for the self-trapping mechanism is in the same order of magnitude as phonon times. Experiments on a shorter timescale, like XPS (X-ray photoelectron spectroscopy) 'see' a static picture of the polaronic particle and detect exotic valence states. Typical behaviour is encountered in WO_3 and structural derivatives containing transition metals such as Mo, Nb, V, Te and Ti. Gehlig *et al.* (1983) and Salje *et al.* (1979) have shown that W^{5+} is the typical ground state in the perovskite lattice including the stabilization due to phonon condensation. Anderson argued (in 1975) that the double occupancy of gap states can be more favourable energetically than in the case of separated electrons, if the electron repulsion is over-compensated by the joint lattice distortion due to both electrons. Consequently, this pair state has been called a bipolaron. Not until 1980 did Schirmer & Salje find conduction bipolarons in low-temperature crystalline WO_{3-x}. These bipolaron states can be dissociated via a photo effect with a maximum efficiency near 1.1 eV, the transport activation energy is 0.18 eV, which is typical for most perovskite structures containing transition metals.

An interesting question arises if we consider perovskite structures with a high carrier concentration. Iguchi *et al.* (1981) predicted theoretically that there exists a maximum carrier density that occurs, preventing any considerable overlap of the polaronic wave functions. If more carriers compete for the same lattice distortion, the surplus carriers have to populate the valence band. This behaviour was found experimentally by Salje & Guttler (1984) in WO_{3-x} where the critical carrier concentration is *ca.* 3.5×10^{21} cm^{-3}. Similar critical densities were also recently observed in structurally closely related $NbO_{2.5-x}$ (Ruscher *et al.* 1988*a*) with $n_c \approx 1.7 \times 10^{21}$ cm^{-1} and W–Nb–O with $n_c \approx 6.10^{21}$ cm^{-1} (Ruscher *et al.* 1988*b*). Structural disorder due to cation disorder on the octahedral site again leads to additional stabilization of the localized states. Electrical transport is then reduced and a Mott-type conductivity ($\ln \sigma \propto T^{\frac{1}{4}}$) is observed in contrast to thermally activated hopping conductivity. The electronic behaviour depends heavily therefore, on the structural state of the materials and varies dramatically if the material is not in thermodynamic equilibrium.

Conclusion

It has been shown that the capacity of the perovskite structure to relax locally under symmetry reduction is the origin for structural phase transitions. It is also the origin for sluggish kinetic behaviour, in particular if impurities play a major role. Possible 'impurities' can be electrons (or holes), which are surrounded by deformation clouds. The electrical conductivity is dramatically reduced by this effect. It is very likely that mantle material of the general composition (Mg, Fe)SiO_3 shows a similar behaviour in thermodynamic equilibrium although such equilibrium is difficult to obtain in laboratory experiments. Any prediction of the behaviour under mantle conditions depends sensitively on the knowledge of the exact chemical composition and the structural state. Here, the effect of Ca and Fe on $MgSiO_3$ is particularly important and might be relevant to our understanding of the nature of the D″ layer. It is also important to note that small changes of the oxygen fugacity can lead to major changes of the physical behaviour of perovskite.

References

Anderson, P. W. 1975 *Phys. Rev. Lett.* **34**, 953–955.
Blinc, R. & Zeks, B. 1979 *Soft modes in ferroelectrics and antiferroelectrics* (ed. E. P. Wohlforth). Amsterdam: North Holland.
Burns, G. & Scott, B. A. 1973*a* *Solid St. Commun.* **13**, 417–421.
Burns, G. & Scott, B. A. 1973*b* *Phys. Rev.* B**7**, 3088–3101.
Cowley, R. A., Buyers, W. J. L. & Dolling, G. 1969 *Solid St. Commun.* **7**, 181–183.
Fleury, P. A., Lazay, P. D. & van Uitot, L. D. 1974 *Phys. Rev. Lett.* **33**, 492–495.
Fleury, P. A., Scott, J. F. & Worlock, J. M. 1968 *Phys. Rev. Lett.* **21**, 16–19.
Gehlig, R., Salje, E., Carley, A. & Roberts, M. W. 1983 *J. solid st. Chem.* **49**, 318–326.
Iguchi, E., Salje, E. & Tilley, R. J. D. 1981 *J. solid st. Chem.* **38**, 342–359.
Kirkpatrick, S. & Sherrington, D. 1978 *Phys. Rev.* B**17**, 4384–4403.
Luthi, B. & Rehwald, W. 1981 *Structural phase transitions I*, (ed. K. A. Muller & H. Thomas). Berlin, Heidelberg, New York: Springer.
Muller, K. A. 1986 *Helv. phys. Acta* **59**, 876–884.
Muller, K. A., Berlinger, W. & Waldner, F. 1968 *Phys. Rev. Lett.* **21**, 814–817.
Ruscher, K., Hussain, A. & Salje, E. 1988*b* *J. Phys.* C. (In the press.)
Ruscher, K., Salje, E. & Hussain, A. 1988*a* *J. Phys.* C **21**, 3737–3749.
Salje, E. 1976*a* *Acta crystallogr.* A**32**, 233–238.
Salje, E. 1976*b* *Ferroelectrics* **12**, 215–218.
Salje, E., Carley, A. & Roberts, W. M. 1979 *J. solid st. Chem.* **29**, 237–251.
Salje, E., Gehlig, R. & Viswanathan, K. 1978 *J. solid st. Chem.* **25**, 239–250.
Salje, E. & Guttler, B. 1984 *Phil. Mag.* B**50**, 607–620.
Salje, E. 1988*a* (ed.) *Physical properties and thermodynamic behaviour of minerals* (ASI 225), pp. 75–118. Dordrecht: Reidel.
Salje, E. 1988*b* *Phys. Chem. Miner.* **15**, 336–348.
Salje, E. & Wruck, B. 1988 *Phys. Chem. Miner.* (In the press.)
Schirmer, O. F. & Salje, E. 1980 *J. Phys.* C **13**, L1067–L1072.
Shirane, G. & Yamada, Y. 1969 *Phys. Rev.* **177**, 858–863.

Discussion

M. F. Osmaston (*The White Cottage, Woking, Surrey, U.K.*). Does Dr Salje think that the high electrical conductivity acquired by perovskite-structured minerals at high impurity concentrations could apply to the D″ layer at the base of the mantle? If so, not only would core–mantle electromagnetic coupling torque become potentially much more significant but the D″ layer might electromagnetically screen us from observing higher frequencies and spatial harmonics in the geomagnetic variation. In view of the importance of removing heat from the core, to sustain convection within it, I wonder what happens to the total thermal conductivity of such minerals under D″ conditions?

E. Salje. I would indeed expect higher electrical conductivity in the D″ layer compared with the mantle material at lower depth. This could also imply strong layering and appreciable gradients of the current densities. It would be very important to obtain experimental results on iron-bearing $MgSiO_3$ under well-defined oxygen fugacities to determine the actual specific conductivities of these materials under mantle conditions.

There is no experimental evidence concerning the thermal conductivity at very high temperatures. As a general rule, we might assume that a high carrier concentration *increases* thermal conductivity. This effect could be compensated, or even over-compensated, by the *decrease* of thermal conductivity due to high defect densities. Iso-spin glasses show often dramatically low thermal conductivity at lower temperatures. It is not clear, however, how to extrapolate these data to higher temperatures and pressures.

Phil. Trans. R. Soc. Lond. A **328**, 417–424 (1989)
Printed in Great Britain

Models of mantle convection: one or several layers

By U. R. Christensen

Max-Planck-Institut für Chemie, Saarstrasse 23, 6500 Mainz, F.R.G.

[Plate 1]

Numerical model calculations are used to determine if convection in the Earth's mantle could be organized in two or more layers with only limited mass exchange in between.

The seismic discontinuity at 670 km depth and the top of the D″-layer at the bottom of the mantle are considered as candidates for internal boundaries. If the 670 km discontinuity is caused by an isochemical phase transition, it has to have a Clapeyron slope of $\mathrm{d}p/\mathrm{d}T \leqslant -6$ MPa k^{-1} to prevent convection currents from crossing; this value is improbably low. If the discontinuity represents a chemical boundary, the intrinsic density difference has to exceed 3 % to prevent subducted lithospheric slabs from penetrating deeply into the lower mantle; also the condition is possibly hard to meet. The least improbable mechanism for a mid-mantle barrier for convection currents would be a combination of endothermic phase transition and chemical change. The boundary between upper and lower mantle would show considerable topography, and a limited material exchange is to be expected at any rate.

The possibility of a downward segregation of former oceanic crust, transformed to dense eclogite, is studied in a further model series. It requires a region of low viscosity, as the D″-layer probably is, and is faciliated by the decrease of the thermal expansion coefficient with pressure. About 20 % of subducted oceanic crust could accumulate at the core–mantle boundary. The dense material would concentrate underneath rising thermal plumes, and some of it is entrained into the plumes, possibly affecting their geochemical signature.

Introduction

Although convection in the Earth's mantle is accepted as a concept, its detailed structure is poorly understood. An outstanding question is the amount of mass exchange between various parts of the mantle. The idea of separately convecting layers above and below the seismic discontinuity at 670 km depth, as opposed to whole-mantle convection, has been discussed for two decades now, without a consensus being reached. The evidence for seismology (Creager & Jordan 1986; Giardini & Woodhouse 1984), from mineral physics (Jeanloz & Thompson 1983; Bukowinski & Wolf 1988) and from geochemistry (Allègre *et al.* 1983; Davies 1984) is ambigious. In this paper, numerical model results are reviewed, which approach this problem from the viewpoint of the mechanism that might enforce the separation of convecting layers.

Recently it was also discussed whether the D″-layer, the lowermost 200 km of the mantle, is chemically distinct from the overlying material (T. H. Jordan, personal communication). Several sources for the origin of such a dense bottom layer are considered: chemical interaction between core and mantle, a relic from the core-formation process, or downward segregation of dense heterogeneities in the convecting mantle. Hofmann & White (1982) proposed for geochemical reasons that subducted oceanic crust, being transformed into dense eclogite, could accumulate at the bottom of the mantle, being stored for about 1–2 Ga, and then rise again in mantle plumes. In the third section, preliminary numerical model results are presented, which indicate that such a process is indeed feasible.

Layering at the 670 km discontinuity

The transformation of the spinel structure of olivine, $(Mg,Fe)_2SiO_4$ into a mixture of perovskite, $(Mg,Fe)SiO_3$, and magnesiowüstite, (Mg,Fe)O, occurs at about the right pressure to explain the second major discontinuity in the mantle. This transition is likely to be endothermic, i.e. it has a negative Clapeyron slope $\mathrm{d}p/\mathrm{d}T < 0$, and it causes an increase of density of the order 6–10%. Some doubt is cast on the interpretation of the 670 km discontinuity as a pure phase boundary by the observation of reflected and converted seismic waves from that depth, implying that the discontinuity is sharp. This is more typically expected for a compositional boundary. A higher magnesium:silicon ratio than in the upper mantle is assumed in most petrological models of a stratified mantle (see Liu 1979). However, the uncertainties in seismic and equation-of-state data are too large to permit a clear decision between uniform or stratified composition. Thus the 670 km discontinuity may represent a pure phase boundary, a compositional boundary, or perhaps a combination of both.

The main influence of a solid–solid transition on convection currents is caused by the deflection of the boundary in hot or cold plumes. For example, in a cold subducted lithospheric slab an exothermic boundary (like the olivine–spinel boundary at 400 km depth) will be elevated to lower pressure because of its positive $\mathrm{d}p/\mathrm{d}T$ slope. The elevated dense phase exerts a strong gravitational force in a downwards direction, which helps to pull down the slab. For an endothermic reaction, a depressed region of the low-density phase occurs, which is buoyant and opposes subduction. If the Clapeyron slope is sufficiently negative, the phase boundary may thus enforce two-layer convection. A further effect of a phase boundary is the release of latent heat. It is important at the critical Rayleigh number (Schubert *et al.* 1975). However, at the high Rayleigh number of mantle convection, its only influence is to modify the adiabat locally. For the question of single- against double-layer convection it is of little consequence (Christensen & Yuen 1985).

Christensen & Yuen (1984, 1985) determined under which conditions a phase boundary would lead to two-layer convection in two-dimensional numerical model experiments. The important parameter is

$$P = \frac{\Delta\rho\gamma}{\rho^2\alpha g h}. \tag{1}$$

Here α is the thermal expansion coefficient, ρ the density, $\Delta\rho$ the density contrast between the two phases, $\gamma = \mathrm{d}p/\mathrm{d}T$ the Clapeyron slope, g the gravity acceleration, and h the height of the mantle. P is a non-dimensional number. Single-layer convection breaks down when P is lower than a certain critical value. Experiments at various values of P indicate that the critical value of P depends slightly on the Rayleigh number Ra (figure 1)

$$P_{\mathrm{crit}} \approx -4.4\, Ra^{-\frac{1}{5}}. \tag{2}$$

Although the circulation pattern was found clearly to be that of two-layer convection for $P < P_{\mathrm{crit}}$, there is still a considerable mass flux across the boundary (figure 1). At a Rayleigh

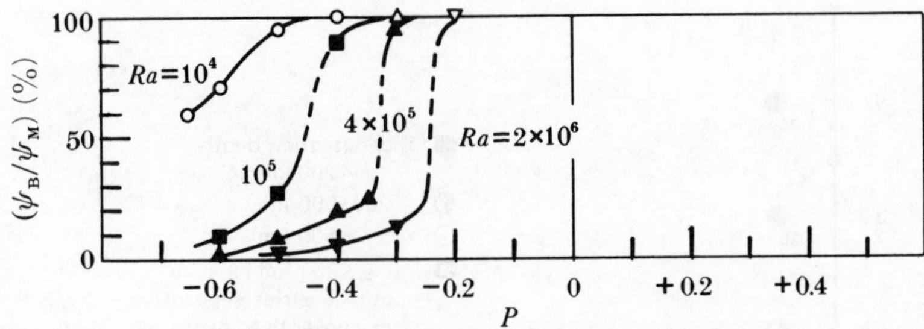

FIGURE 1. Results from convection models in a square box with a phase boundary at mid-depth. The ratio of the stream-function maximum at the phase boundary to the absolute maximum is plotted against the parameter P (equation (1)); it is a measure for the degree of mass exchange between both layers.

number appropriate for present-day mantle convection, say 10^7, $P_{crit} \approx -0.18$. Assuming plausible values for the various parameters that enter into equation (1), one finds that dp/dT has to be less than about -6 MPa K^{-1} for the spinel → perovskite + magnesiowüstite transition to cause layered convection. Preliminary experimental values are only of the order of -2 MPa K^{-1} (Ito & Yamada 1982). Therefore it seems unlikely that the phase change would lead to layered convection unless future experiments would indicate a much lower value of dp/dT. It is interesting to note that for the large icy satellites of the outer planets the story could be different. For the ice-II–ice-V transition in Ganymede, $P = -1.6$, thus the critical value is by far exceeded and two-layered convection must be expected if the temperature is in the stability field of these phases.

Christensen & Yuen (1984) also considered the influence of a chemical or a combined chemical and phase boundary in numerical convection models including subducted lithospheric slabs. The results are summarized in a domain diagram (figure 2) for the style of convection. We estimated the uncertainties in the exact limits of the domains to be on the order of some tens of percent, because of uncertainties in parameter values and unmodelled effects, for example the influence of the driving buoyancy of the olivine–spinel transition. According to our model results, the slab would be stopped at the boundary and bend sideways when the compositional density difference exceeds about 4–5%. The impact of the slab causes a depression of the boundary between upper and lower mantle of 100–200 km. It is not very likely that the compositional density difference can exceed a few percent. When it is less than 4% the slab was found to plunge deeply into the lower mantle and with less than 2% it would reach the core mantle boundary and lead to large-scale mixing of upper- and lower-mantle material. Kincaid & Olson (1987) found, at least semi-quantitatively, the same behaviour in laboratory experiments using corn syrup to model slabs in a stratified convecting system.

The least improbable mechanism for two-layer convection (in a narrow sense) is a combination of a distinctly negative dp/dT of the spinel → perovskite + magnesiowüstite transition and a superimposed compositional change. As can be seen from figure 2, 3% chemical density difference and a Clapeyron slope of -3 MPa K^{-1} would lead to layered convection. Both are still plausible values. Of course, having a chemical change and a phase boundary at the same depth must appear as an improbable coincidence, unless someone can put forward a plausible explanation for the development of such a situation, for example within the scenario of an early magma ocean. The depression of the boundary between upper and lower mantle at the site of slab subduction would be of the order of 100 km. It should be possible to detect such a depression by seismological means. However, one would expect travel-

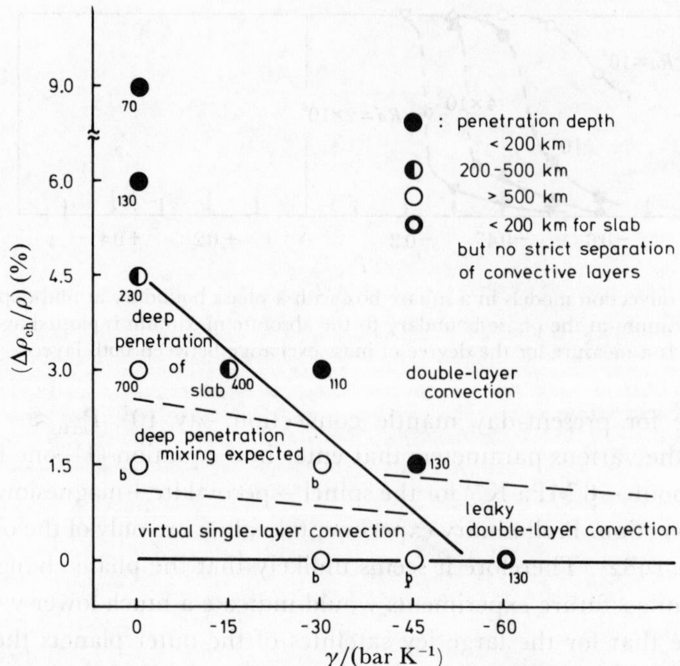

FIGURE 2. Domain diagram for convection including subducting plates. It indicates the mode of convection depending on the chemical contribution to the density differences at the 670 km discontinuity and the p/T-slope of the phase transition contribution. Numbers indicate the penetration depth of slabs below 670 km, 'b' means penetration to the bottom of the mantle. $(\Delta\rho/\rho)_{tot} = (\Delta\rho/\rho)_{ph} + (\Delta\rho/\rho)_{ch} = 9\%$. 1 bar = 10^5 Pa.

time anomalies from deep-focus earthquake that show more or less the opposite sign to what has been found (Creager & Jordan 1986).

In summary, from the point of view of the mechanism for dynamical layering, whole-mantle convection is preferred. On the other hand, the conditions for having separate convection layers above and below the 670 km discontinuity are not so extreme that this possibility could be ruled out entirely. Further evidence could come from mapping the boundary in the region of slab subduction. It should be emphasized that, whatever the mechanism for possible layering, the mass exchange between upper and lower mantle would still be significant.

Subducted crust and a chemical D″-layer

Recently it has been speculated that the D″-layer at the bottom of the mantle is a chemical boundary layer (T. H. Jordan, personal communication) to explain the strong seismic lateral heterogeneity in this region and the inferred topography of the core–mantle boundary, which is on the order of ± 6 km (Morelli & Dziewonski 1987). Among the various proposed mechanisms for its formation, perhaps the least speculative is the segregation of subducted oceanic crust. At least we know that large amounts of basaltic crust are actually subducted and that it transforms under pressure into eclogite and other dense phases. According to Ringwood (1982) the former crust keeps an excess density of about 5% compared with normal mantle through all the phase transitions that occur on the way to the lower mantle. Recent work (Irifune & Ringwood 1987) indicates that it may be buoyant in a limited depth interval around 670 km; however, if slabs can penetrate below this zone it would be of little importance for the processes considered here. An obstacle for segregation is that the crust lies above a layer of depleted harzburgite, which is compositionally the complement to the crust and less dense than normal mantle by 1–2%. Both layers combined are about neutrally buoyant (except for thermal buoyancy). Another obstacle is that the crustal layer is thin, about 6–7 km. Stokes-sphere estimates show that a body of eclogite of 5 km diameter would sink only at 0.1 mm a^{-1} for a typical viscosity value of the mantle of 10^{21} Pa s. This is much less than the expected velocity of convection, and mixing rather than segregation would be expected. Model calculations with constant viscosity by Gurnis (1986) indeed show for realistic parameters not much of a settling effect for heavy tracer particles inserted into a convective flow. However, because the viscosity is strongly temperature dependent, localized regions exist in the mantle where the viscosity is much below its mean value, namely the hot lower boundary layer of convection, or hot rising plumes. Here the separation of 'crust' and 'harzburgite' may take place with the former accumulating at the bottom.

A number of numerical model experiments have been performed to determine whether such a process is indeed feasible with realistic parameters. Into time-dependent models of variable viscosity convection, a chemically layered 'lithosphere' is inserted at time zero, consisting of a dense thin crust and a thicker buoyant layer underneath. Both combined are neutrally buoyant. Technically the compositional differences are modelled by statistically distributed tracer particles. To avoid spurious effects, their number must be so large that the influence of a single tracer is negligible and only their statistical density effects the flow. Up to 10000 tracer particles have been used. The effective Rayleigh number is of the order of 3×10^5, about 1 or 2 orders of magnitude lower than what is realistic for whole-mantle convection. This poses the question of how to scale various parameters to get similar behaviour as in the mantle. The two conditions, which are met in the calculations, are that the ratio of chemical to thermal density anomalies is the same as for the Earth

$$\Delta\rho_{\text{chem}}/\rho\alpha\Delta T \approx 1, \tag{3}$$

and that the ratio of the thickness of the chemical boundary layer (the oceanic crust) to the thermal boundary layer (the lithosphere) is correct:

$$\delta_{\text{chem}}/\delta_{\text{therm}} \approx 0.1. \tag{4}$$

A last point is how to translate model time into real time. Here a timescale based on the mean flow velocity is taken. The time that a particle needs to move a distance equal to the depth of the convection layer is called the transit time. For the Earth, one transit time is about 60 Ma. This figure is used to translate the time lapse in the model into geological time.

The standard parameters in the model comprise an exponential decrease of the thermal expansion coefficient with depth by a factor of three (Stacey 1977), a decrease of the viscosity due to its temperature dependence from top to bottom by a factor of 16000, and an increase due to pressure by a factor of 64. This combines into a viscosity difference between top and bottom by a factor of 250. No internal heating is considered and all boundaries are stress-free. A model run with these parameters in a box of aspect ratio 1.5 shows in fact a strong separation effect in the lower boundary layer (figure 3, plate 1). The former harzburgite layer rises in a buoyant chemical diapir, and the former crust accumulates in a dense pool underneath the rising thermal plume. This pool is not entirely crust, it contains also some normal mantle and a little harzburgite and has about half the excess density of the crust. The pool is leaking;

continuously some material is entrained into the thermal plume. However, even after 18 transit times (1 Ga) about half of the original crustal material resides in the pool.

This model experiment is unrealistic in the sense that subduction must occur at the impenetrable sidewall of the box. This guides the lithosphere directly into the hot low-viscosity bottom layer, where it lies upside down, which highly favours the separation effect. However, before considering a more realistic model, the influence of two variations of the standard parameters were studied. When the thermal expansion coefficient is not depth-dependent, but set to a mean value, again a pool of segregated crust forms. However, this time the pool is leaking much more heavily and has lost most of its substance after seven transit times (400 Ma). In another experiment, the viscosity, not the expansion coefficient, was fixed at a mean value. This time no separation at all was found; rather the compositional layers stayed together and were almost passively torn around by the flow. This emphasizes the importance of the local viscosity reduction due to temperature dependence. For example, when a pool has formed, the low viscosity reduces the shear forces of the external flow that act on the pool such as to drag out material.

To avoid the unrealistic geometrical constraint on subduction at a sidewall of the computational box, model runs have been done in a domain of aspect ratio three with two cells (initially), where subduction takes place in the middle. In this configuration the flow is strongly time-dependent due to boundary instabilities and the subducting chemically layered lithospheres does not go straight to the bottom but is first turned sideways, although eventually some of its material enters the lower thermal boundary layer. This time only 5% of the subducted crust was found to form a tiny dense pool under one of the thermal plumes, whereas the rest was scattered throughout the cells. This seems discouraging; however, in the model a crustal layer was only once subducted, whereas on the Earth crustal subduction is a continuous process and new crust could be added to pre-existing dense pools.

The presence of previously segregated material at the bottom seems also to enhance the addition of new crust, although the reason for this is not clear. In two model runs that were started with a compositionally stratified lithosphere and some dense material already at the bottom, about 20–25% of the subducted crust was found to join into the bottom pools. In figure 4, plate 1, it can be seen how part of the subducted crust merges with the pre-existing dense pool.

Further model experiments were concerned with the dynamics rather than with the formation of a chemical bottom boundary layer. They were started by putting dense material right onto the bottom. It was found that the pools (which quickly formed) can wander around and follow the upwelling currents. They can also split up and merge again, analogous to their

---

Description of plate 1

Figure 3. Four snapshots of the temporal evolution of a convection model starting with a chemically stratified top layer with dense crust in red (dark shading) and buoyant harzburgite in blue (faint shading). The instantaneous stream-lines are drawn, numbers at the margin indicate the time lapse in transit time units.

Figure 4. Two instances in time of a convection model with subduction initiated in the center of the model box with a chemically stratified top layer and an already existing dense bottom layer. At transit time 2.5 the subducted former lithosphere is torn and distorted. Some of the former crust merges with the dense pool on the left side. At time 4.1 two distinct pools, sitting underneath rising currents are found; they have consumed 25% of the subducted crust and they leak into the rising plumes.

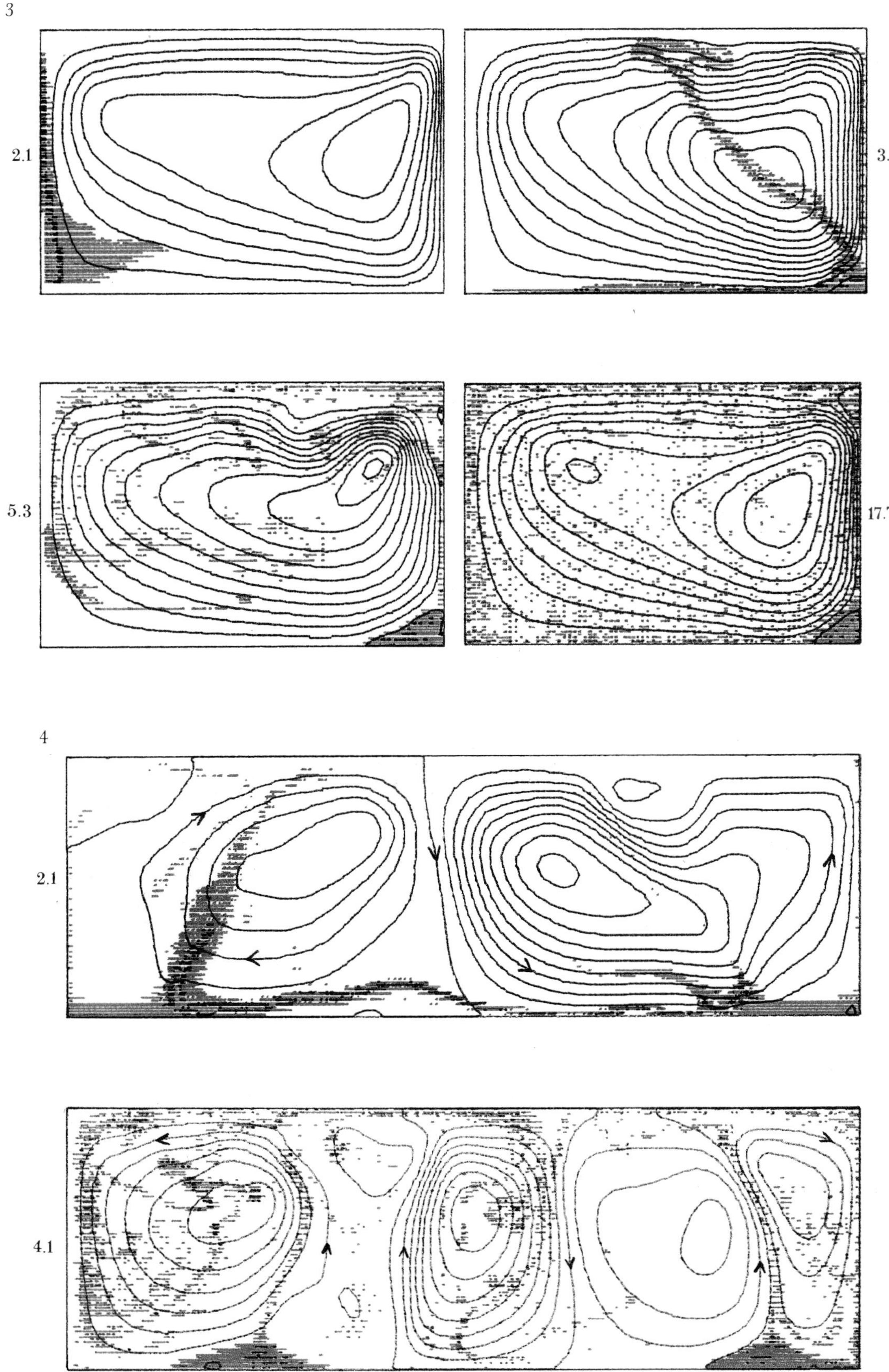

FIGURES 3 AND 4. For description see opposite.

counterpart at the Earth's surface: the continents. They show internal circulation with the opposite sense of rotation to the external flow. They do not only leak into the rising plumes, but they also entrain ambient mantle. Thus they become progressively diluted and destabilized and finally dissolve rapidly after about 1 Gȧ. In the real Earth there may, however, be a dynamical equilibrium between dilution by entrainment, loss by leaking into plumes, and replenishment by the addition of new subducted crust. In a few examples the bottom topography has been calculated. As would be expected there is a local low under the dense pools. However, this depression is not larger in amplitude than the variation that is typically produced by thermal convection alone (see also Davies & Gurnis 1986). This can be understood in the following way: the topographic depression is more or less related to the height of the dense pool. Pools form because the dense material is swept laterally by thermal convection currents. How high the material can pile up depends only on the strength of thermal convection; without it the pools would flow flat like pancakes, especially in the low viscosity of the bottom layer. Thus the topographic variation of the bottom depends in an indirect way on the thermally driven flow.

Conclusions

Two kinds of layering in mantle convection have been studied in numerical models, considering the mechanisms for stabilization or formation of such layers. Separate convection in the upper and lower mantle, divided by the 670 km discontinuity, could be caused by a very endothermic nature of the spinel $\rightarrow$ perovskite + magnesiowüstite transition, by a gross difference in the bulk chemistry of upper and lower mantle, or a combination of both. However, the required parameter values are somewhat extreme and with our present knowledge it is not very likely that the conditions are met. Whole-mantle convection is the preferred mode of convection. This does not exclude the possibility that the mass exchange between upper and lower mantle could be lower than previously assumed, for example because of high viscosity in the lower mantle (Davies 1984).

The model results are more favourable for another kind of hypothetical 'layering', the formation of mantle dregs in the D″-layer, where former subducted oceanic crust could partially accumulate. The model calculations show the great importance of temperature-dependent viscosity for the segregation, and, to a lesser extent, of the decrease of the thermal expansion coefficient with depth. The ancient crust would form local pools underneath rising thermal currents (mantle plumes), which entrain some of the dense material. A sufficiently large amount of dense material would form a continuous cover around the core. Only in this case it would thermally insulate the core to a significant degree, but not if part of the core–mantle boundary were swept clean of it. The model calculations presented here are somewhat preliminary and leave a number of questions. It is necessary to expand the parameter range to define more clearly the conditions for the segregation effect. A number of important points can, in principle, be addressed in numerical calculations, for example the question of the average residence time of material in D″, or if the amount of material entrained into plumes would be large enough to affect significantly the geochemical signature of hot-spot basalts. Entrainment of significant amounts of ancient oceanic crust could explain some, but not all, of the geochemical differences between MORBS (mid-ocean ridge basalts) and hot-spot basalts. The high $^{3}He/^{4}He$ ratios of some oceanic islands warrant another explanation. The influence on core–mantle topography and heat flow, and the effect on seismic travel times could be systematically investigated in numerical models. The present model calculations have indicated that the segregation process can operate and leave it to future work to quantify the details.

References

Allègre, C. J., Staudacher, T., Sarda, P. & Kurz, M. 1983 *Nature, Lond.* **303**, 762–766.
Bukowinski, M. S. T. & Wolf, G. H. 1988 *J. geophys. Res.* (In the press.)
Davies, G. F. 1984 *J. geophys. Res.* **89**, 6017–6040.
Davies, G. F. & Gurnis, M. 1986 *Geophys. Res. Lett.* **13**, 1517–1520.
Creager, K. C. & Jordan, T. H. 1986 *J. geophys. Res.* **91**, 3573–3589.
Christensen, U. & Yuen, D. A. 1984 *J. geophys. Res.* **89**, 4389–4402.
Christensen, U. & Yuen, D. A. 1985 *J. geophys Res.* **90**, 10291–10300.
Giardini, D. & Woodhouse, J. H. 1984 *Nature, Lond.* **307**, 505–509
Gurnis, M. 1986 *J. geophys. Res.* **91**, 11407–11419.
Hofmann, A. W. & White, W. M. 1982 *Earth planet. Sci. Lett.* **57**, 421–436.
Liu, L. G. 1979 *Earth planet. Sci. Lett.* **62**, 91–103.
Ito, E. & Yamada, H. 1982 *High pressure research in geophysics*, pp. 405–419. Tokyo Center for Academic Publishing.
Irifune, T. & Ringwood, A. E. 1987 *Earth planet. Sci. Lett.* **86**, 365–376.
Jeanloz, R. & Thompson, A. B. 1983 *Rev. Geophys.* **21**, 51–74.
Kincaid, C. & Olson, P. 1987 *J. geophys. Res.* **92**, 13832–13840.
Morelli, A. & Dziewonski, A. M. 1987 *Nature, Lond.* **325**, 678–683.
Schubert, G., Yuen, D. A. & Turcotte, D. L. 1975 *Geophys. Jl R. astr. Soc.* **42**, 705–735.
Ringwood, A. E. 1982 *J. Geol.* **90**, 611–643.
Stacey, F. 1977 *Physics Earth planet. Inter.* **15**, 341–348.

Phil. Trans. R. Soc. Lond. A **328**, 425–439 (1989)
Printed in Great Britain

Geochemistry and models of mantle circulation

By A. W. Hofmann
Max-Planck-Institut für Chemie, Postfach 3060, 6500 *Mainz, F.R.G.*

Geochemical data help to constrain the sizes of identifiable reservoirs within the framework of models of layered or whole-mantle circulation, and they identify the sources of the circulating heterogeneities as mainly crustal and/or lithospheric, but they do not decisively distinguish between different types of circulation.

The mass balance between crust, depleted mantle and undepleted mantle based on $^{143}Nd/^{144}Nd$, Nb/U and Ce/Pb, and the concentrations of very highly incompatible elements Ba, Rb, Th, U, and K, shows that *ca.* 25–70% (by mass) of depleted mantle balances the trace element and isotopic abundances of the continental crust. This mass balance reflects the actual proportions of mantle reservoirs only if there are no additional unidentified reservoirs.

Evidence on the nature and ages of different source reservoirs comes from the geochemical fingerprints of basalts extruded at mid-ocean ridges and oceanic islands. Consideration of Nd and He isotopes *alone* indicates that ocean island basalts (OIBs) may be derived from a relatively undepleted portion of the mantle. This has in the past provided a geochemical rationale for a two-layer model consisting of an upper depleted and a lower undepleted ('primitive') mantle layer. However, Pb-isotopic ratios, and Nb/U and Ce/Pb concentration ratios demonstrate that most or all OIB source reservoirs are definitely *not* primitive. Models consistent with this evidence postulate recycling of oceanic crust and lithosphere or subcontinental lithosphere. Recycling is a natural consequence of mantle convection. This cannot be said for some other models such as those requiring large-scale vertical metasomatism beneath OIB source regions.

Unlike other trace elements, Nb, Ta, and Pb discriminate sharply between continental and oceanic crust-forming processes. Because of this, the primitive mantle value of Nb/U = 30 (Ce/Pb = 9) has been fractionated into a continental crustal Nb/U = 12 (Ce/Pb = 4) and a residual-mantle (MORB (mid-ocean ridge basalt) plus OIB source) Nb/U = 47 (Ce/Pb = 25). These residual mantle values are uniform within about 20% and are not fractionated during formation of oceanic crust. By using these concentrations ratios as tracers, it can be shown that the possible contribution of recycled continental crust to OIB sources is limited to a few percent. Therefore, recycling must be dominated by oceanic crust and lithosphere, or by subcontinental lithosphere. Oceanic crust normally bears a thin layer of pelagic sediment at the time of subduction, and this is consistent with OIB sources that are dominated by subducted oceanic crust with variable but always small additions of continental material.

Primordial ^{3}He, ^{36}Ar, and excess ^{129}Xe, in oceanic basalts demonstrate that the mantle has been neither completely outgassed nor homogenized, but they do not constrain the degree of mixing or the size of reservoirs. Also, helium does not correlate well with other isotopic data and may have migrated into the basalt source from other regions. The high $^{3}He/^{4}He$ ratios found in some OIBs suggest that, even though the basalts are not derived from primordial mantle, their sources may be located close to a reservoir rich in primordial gases. This leads to models in which the OIB sources are in a boundary layer within the mantle. The primordial helium migrates into this layer from below. The interpretation of the rare-gas data is still quite controversial.

It is often argued that the upper mantle is a well-homogenized reservoir, but the data indicate heterogeneities on scales ranging from 10^0 to 10^6 m. The $^{206}Pb/^{204}Pb$ ratios in the oceanic mantle range from 17 to 21, which is similar to the range in most continental rocks. The degree of mixing cannot be directly inferred from these data unless the size and composition of the heterogeneities and the time of their introduction into the system are known. The relative uniformity of Nb/U and Ce/Pb ratios in the otherwise heterogeneous MORB and OIB sources indicates that this reservoir was indeed homogenized after the separation of the continental crust, and that the observed isotopic and chemical heterogeneities were introduced subsequently.

Overall, the results are consistent with, but do not prove, a layered mantle where the upper layer contains both MORB and OIB sources, and the lower, primitive mantle is not sampled by present-day volcanism. Alternative models such as those involving a chemically graded mantle have not been sufficiently explored.

Introduction

Studying mantle geochemistry is somewhat like examining the surface of a marble cake and trying to deduce its internal structure from the observed distribution of brown and yellow. Mantle tomography offers the hope for 'seeing' the internal structure and thereby interpreting the marble cake. But because we are working with tracers that do not affect the physical properties of the rock in any measurable and direct way, this effort will have to depend on interpreting *correlations* of velocity structure and geochemical tracers.

The more traditional, although usually not explicitly stated, approach is to interpret the geochemical surface map using what is known about plate kinematics and, to a lesser extent, convection dynamics. For example, plumes are more likely to be formed from low-density boundary layers than from an internally heated medium. The plumes are stationary or at least mechanically decoupled from the plate movements and their sources are therefore believed to be located deeper in the mantle. In contrast, ridge migration is strongly coupled to the geometry and movement of the plates and is 'passive' in this sense. Therefore, the ridges sample predominantly the uppermost mantle, not a rising deep-mantle current. Without such additional constraints or inferences, geochemistry alone has very little to say about mantle structure.

Geochemical tracers of mantle reservoirs

The conventional use of geochemical tracers to characterize source regions of basaltic melts in the mantle rests on the assumption that the isotopic composition of a partial melt in the mantle is representative of the entire volume of mantle rock from which the melt has been extracted. This assumption has been questioned repeatedly during the past 15 years. Hofmann & Hart (1978) argued on the basis of measured and estimated diffusion coefficients in solid and liquid silicates that partially molten mantle rocks should come to local equilibrium between solid and melt in geologically very short times of the order of less than 10^4 years. They argued further that if this were not so, then much of the work done in experimental petrology on the melting relations of mantle rocks would be meaningless. Most workers subsequently accepted the conclusion that individual mineral grains are extremely unlikely to survive in a state of disequilibrium with partial melts except in situations where melts are formed and extracted very rapidly. These conclusions have been reaffirmed by Hart & Zindler (1988) in the light of more recent experimental data. Examples of exceptions might be found in partially molten xenoliths or perhaps in the wall rock of a rapidly rising hotter diapir. Ordinary convective motions in the mantle should always be slow enough to ensure attainment of local equilibrium, especially so because the pressure and temperature changes accompanying these motions force the material to recrystallize as it undergoes phase changes.

Mesoscale heterogeneities

It is appropriate to ask the question how far in space does local equilibrium extend. What happens on the decimetre to kilometre scale where diffusional transport becomes negligibly small, even in partially molten systems and over geologically long periods of time? When mantle peridotites are examined in outcrop, it becomes immediately clear that these rocks can be quite heterogenous on this scale (Reisberg & Zindler 1986). Whether the heterogeneities are veins (Hanson 1977), trace-element enriched lumps (Sleep 1984), dikes or (almost) infinitely drawn out former oceanic crust (Allègre & Turcotte 1986), if the 'enriched' portions can be melted preferentially and this melt be extracted without chemical and isotopic equilibration with the more 'depleted', refractory portions, then the isotopic composition of the melt will be biased in favour of the enriched portions of the source. This theme has been discussed in numerous variants (Hanson 1977; Sleep 1984; Hart & Zindler 1986; Zindler *et al.* 1984; Allègre & Turcotte 1986; Fitton & Dunlop 1985). The consequence common to all these variants is that the isotopic composition of the melt will depend on the overall degree of melting, because this determines how much of the more-refractory portion of the source rock is sampled by the melt. For the sake of convenience, I will call the scale ranging from decimetre to kilometre the *mesoscale*.

Whether or not the partially molten mantle attains equilibrium on the mesoscale depends on whether or not the melt permeates all the refractory portions of the source region. If it does, diffusion in the melt alone will tend to equilibrate the melt over a distance of a metre in a few thousand years. Migration and mixing of the melt will greatly enhance the overall equilibration, and it is likely that this sort of equilibrium will effectively extend to the kilometre scale. If the melt does *not* permeate all of the rock, isotopic and chemical equilibration is quite unlikely on the mesoscale, because the solids that are not in direct contact with an intergranular melt (or fluid) are unlikely to equilibrate with it. Experiments by Waff & Buhlau (1979) have shown that a partial melt does not wet the intergranular surfaces, but it does penetrate peridotite through a fully interconnecting network porosity along grain corners. If this is also true in the mantle, then mesoscale equilibrium should easily be attained.

Another approach to this problem is to study melts and residues in natural settings where the degree of melting changes with time or location. If mesoscale disequilibrium prevails, then those regions that have undergone the lowest degree of melting should have produced melts with the isotopic characteristics of the highest incompatible-element enrichment, because these melts would preferentially sample the most fertile and most trace-element enriched portions of the source. Several cases have been documented where the opposite is true. Hawaiian lavas consistently change to lower $^{87}Sr/^{86}Sr$ and higher $^{143}Nd/^{144}Nd$ ratios as the degree of melting decreases toward the end of the life of each volcano (Chen & Frey 1985; Feigenson 1984; Lanphere *et al.* 1980; Hofmann *et al.* 1987). Along the mid-Atlantic ridge, the most enriched melts are found in regions where the residual periodites indicate that the largest melt fraction has been extracted. Small sea mounts sometimes show very heterogeneous compositions

(Zindler *et al.* 1984), which may well be caused by mesoscale disequilibrium, but most show the same depleted isotopic and chemical character as nearby ordinary ridge segments (Batiza & Vanko 1984). Thus, although mesoscale disequilibrium cannot be dismissed out of hand, the weight of the evidence points to a rather minor role at best, and there are good reasons to think that the regional variations displayed by oceanic basalts do indeed reflect regional, rather than mesoscale heterogeneities in the mantle. The applicability of geochemical tracers to problems of large-scale mantle structure and circulation depends on the correctness of this inference.

Geochemical tracers

(a) *Isotopes*

Isotopic compositions of heavy, refractory elements are the most reliable tracers of the source compositions of mantle derived basalts. This is so because any mass-dependent isotopic fractionation during melting is negligible. Even if such fractionation did occur in Nature, it would be automatically corrected during mass spectrometric analysis. Non-mass-dependent fractionation, as suggested by O'Hara & Mathews (1981), has never been observed in the multi-isotopic elements Sr and Nd where, purely for reasons of analytical quality control, two or more non-radiogenic isotopic ratios are routinely monitored for non-mass-dependent effects in many laboratories. These are trivial matters to isotope geochemists, but they must be mentioned periodically because they are still raised frequently in lectures and discussions.

The chief disadvantage of isotope ratios is that they are applicable to only a small number of elements that contain radiogenic daughter isotopes of long-lived parent nuclides. Three of the most commonly used isotopic tracers, Sr, Nd, and Hf, are strongly correlated in most mantle derived rocks (Richard *et al.* 1976; DePaolo & Wasserburg 1976; O'Nions *et al.* 1977; Patchett & Tatsumoto 1980) and do not yield much independent information. The resulting so-called mantle array has been variously interpreted as mixtures of primitive and depleted sources, mixtures of continental or oceanic crust and depleted mantle, and mixtures of metasomatically enriched and depleted mantle. Because of this, they are very useful for the purpose of constraining mass balances of mantle reservoirs and mapping the isotopic heterogeneities at the surface, but not for identifying the nature of the source materials. Deviations from the purely linear correlation in several ocean island basalts (OIBS) (White & Hofmann 1982) demonstrate that more than two source reservoirs must be present, but this alone does not constrain the origin of these reservoirs. It seems unlikely that additional data on Sr, Nd, and Hf isotopes will change this. The continental crust has been extracted from the mantle and is enriched in the so-called incompatible elements. If other enrichment processes have operated in the mantle, they may have created parent–daughter fractionations that are indistinguishable from the differentiation of primitive mantle into continental crust and depleted residue.

The isotopic ratios of Pb and He are well correlated with those of Sr in some regions but not in others, and their interpretation is still being disputed by the specialists. Reviews of the combined isotopic data in OIB and mid-ocean ridge basalts (MORBS) have led to the tentative identification of as many as five distinct source reservoirs (White 1985; Zindler & Hart 1986), but it is far from clear just how separate these sources are and how they may have been produced. Some of the issues involved will be discussed further below.

(*b*) *Trace elements*

Trace element abundance ratios are suitable tracers under certain circumstances, where they can be used as 'quasi-isotopic' ratios. They considerably expand the geochemical arsenal available to fingerprint and identify sources, and some of these ratios permit much more specific source assignments than is possible on the basis of isotopic ratios alone. However, there is always the risk of trace-element fractionation between the basalt sampled at the surface and its source in the mantle. This fractionation is minimized when the melt fraction is much greater than the bulk partition coefficients of the elements involved. This is illustrated in figure 1, which is a plot of the equation

$$\frac{c_1}{c_2} = \frac{c_1^0}{c_2^0} \cdot \frac{F + D_2(1-F)}{F + D_1(1-F)}, \tag{1}$$

where c and c^0 are the concentrations in the melt and in the source, respectively, F is the melt fraction and D is the bulk partition coefficient defined as $D = c(\text{solid})/c(\text{melt})$. The subscripts refer to elements 1 and 2. For example, the widely used Zr/Nb or La/Sm ratios may well be perfectly good tracers in MORBS because the melt fractions are comparatively large, but not in some OIBS, where melt fractions are smaller and the same elements may be fractionated.

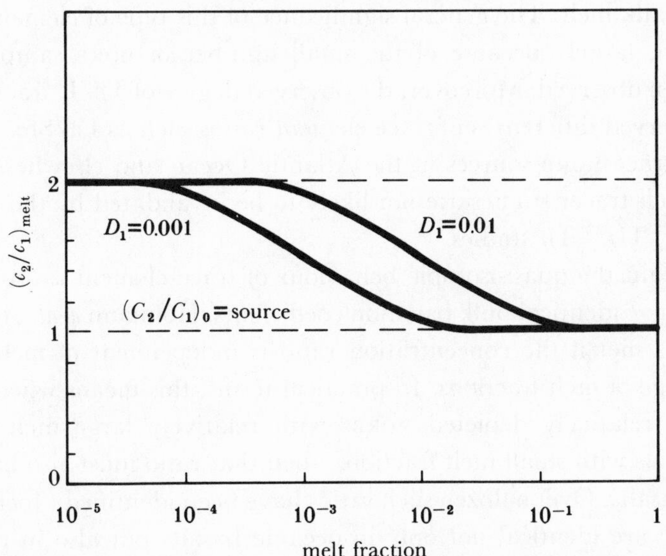

FIGURE 1. Concentration ratios of two trace elements in a melt as a function of melt fraction (equation (1)). The two curves are for bulk partition coefficients between solid and melt, $D_1 = 0.001$ and $D_1 = 0.01$. The ratio of bulk partition coefficients is in both cases $D_2/D_1 = 2$. For highly incompatible elements, ($D < 0.001$), and melt fractions, $F > 0.01$, the concentration ratio in melt approaches the source ratio. This is the basis for using trace-element concentration ratios to characterize source compositions.

U and Th are among the most incompatible elements in mantle rocks, that is, their bulk partition coefficients between solid and melt are very small, much smaller than melt fractions believed to be realistic for MORBS. Yet, the thorium isotope ^{230}Th in many MORBS is not in radioactive equilibrium with its parent nuclide ^{238}U (Allègre & Condomines 1982). Such a disequilibrium in a radioactive decay chain is possible only if there has been a chemical fractionation (or separation) between parent and daughter within the geologically very short time span of several half lives of ^{230}Th ($t_{\frac{1}{2}} = 7.5 \times 10^4$ years). The most likely, although unexpected and somewhat alarming, explanation of this phenomenon is that Th is fractionated from U between the MORB and its source (Allègre & Condomines 1982), and this raises serious doubts about the 'quasi-isotopic' status of other abundance ratios of highly incompatible elements. U and Th have partition coefficients in the major mantle minerals olivine, orthopyroxene, clinopyroxene, garnet, and spinel of $D < 10^{-2}$ (Watson *et al.* 1987), and the bulk partition coefficients of mantle mineral assemblages consisting of these phases (but dominated by olivine and orthopyroxene) should have bulk partition coefficients of $D < 10^{-3}$. Figure 1 shows that such elements should not be fractionated in MORBS, which have melt fractions far exceeding the critical value of $F = 0.01$ (see Klein & Langmuir 1987; Hofmann 1988). Galer & O'Nions (1986) explained this unexpected fractionation as follows: in the unmelted mantle, the highly incompatible elements are concentrated in trace minerals, which have comparatively large partition coefficients for these elements. During mantle upwelling beneath the ocean ridges, partial melting at the base of the melting regions begins necessarily at very small melt fractions, so that the trace-mineral phases are not immediately consumed by the melt. They can therefore fractionate these elements, and if the melt fraction is able to migrate upward slightly faster than the solid phases, the fractionated melt enters the higher regions where melt fractions become naturally larger. The trace-element-rich melt derived from below now mixes with the locally produced melt and imposes its inherited trace element composition on the bulk melt. The general significance of this type of element fractionation is still rather uncertain, largely because of the small number of MORB samples in which the phenomenon has been observed. Moreover, the observed degree of Th–U fractionation is much smaller than the observed differences of trace element ratios such as La/Sm, which have been used extensively to trace MORB sources in the Atlantic Ocean and elsewhere (Schilling *et al.* 1983). Therefore, such tracer studies are not likely to be invalidated by the evidence derived from $^{230}Th/^{232}Th - ^{238}U/^{232}Th$ studies.

One way to ascertain the quasi-isotopic behaviour of trace-element ratios is to choose two trace elements that have identical bulk partition coefficients. Hofmann *et al.* (1986) have shown that this condition is met if the concentration ratio is independent of melt fraction over a sufficiently large range of melt fractions. In practical terms, this means when a trace element ratio is identical in relatively depleted MORBS with relatively large melt fractions *and* in relatively enriched OIBS with small melt fractions, then that ratio must also be the same in the sources of all these basalts. Over a dozen such ratios have been identified (Jochum & Hofmann 1989). Most of these are identical not only in oceanic basalts but also in most rocks of the continental crust. They are therefore not useful as tracers of terrestrial fractionation processes but merely confirm the meteorite model for the composition of the Earth. However, two of these ratios, Nb/U and Ce/Pb, are higher by a factor of four in the mantle sampled by MORBS and OIBS than in the continental crust. These ratios are unique tracers that can distinguish between mantle enrichments which have produced OIB sources and sources derived from the continental crust.

Mass balance

Some of the earliest studies of ocean-floor basalts revealed that they have unusual chemical and isotopic compositions. Beginning with Tatsumoto *et al.* (1965) who showed that MORBS have very low $^{87}Sr/^{86}Sr$ ratios, and K, Rb, U and Th concentrations, these rocks were subsequently found to be also relatively depleted in Cs, Ba, Nb, Ta, and the light rare-earth elements. It also became clear that, qualitatively, their chemistry is complementary to that of the continental crust. Those elements that are most strongly enriched in the continental crust are also the most incompatible in the mantle, that is, they are partitioned most strongly into any available silicate melt in contact with the common minerals of the upper mantle, olivine, pyroxene, garnet and spinel. From an *a priori* knowledge of the bulk earth composition (derived from chondritic meteorites with some important modifications particularly to volatile elements), Taylor & McLennan (1985, pp. 266, 267) estimated that about 30% of the total terrestrial inventory of these 'incompatible' elements are now in the continental crust.

The relation of the continental crust to the source reservoir of ocean floor basalts was quantified when isotopic data for neodymium became available. The $^{143}Nd/^{144}Nd$ ratio of the bulk earth is identical to that of chondritic meteorites. Moreover, this ratio is relatively uniform in most continental rocks and also in MORBS; consequently, a rather well constrained mass balance can be calculated using the simple assumption that primitive mantle material has been differentiated into ('enriched') continental crust and complementary ('depleted') mantle (DePaolo 1980; Jacobsen & Wasserburg 1979; O'Nions *et al.* 1979; Allègre *et al.* 1983.) The simplest expression of this has been formulated in the following three equations (Davies 1981):

$$m_p + m_d + m_c = 1, \tag{2}$$

$$m_d C_d \epsilon_d + m_c C_c \epsilon_c = 0, \tag{3}$$

$$m_p C_p + m_d C_d + m_c C_c = C_p, \tag{4}$$

where m are the mass fractions of the three reservoirs p (primitive mantle), d (depleted mantle), and c (continental crust), C are the respective neodymium concentrations, and ϵ are the relative deviations of the $^{143}Nd/^{144}Nd$ ratios from the chrondritic (primitive mantle) value, which is defined by $\epsilon_p \equiv 0$. The ϵ values, the mass of the continental crust, and the Nd concentrations of the primitive mantle and the continental crust are measured or assumed, so that the equations can be solved for m_p, m_d, and C_d. These calculations yielded the important result that the depleted fraction of the mantle $m_d \approx 0.25$–0.5. Davies (1981) and Allègre *et al.* (1983) estimated that the errors in the input parameters are large enough so that m_d might be as high as 0.8. Recent re-evaluations of the neodymium concentrations of the bulk silicate earth (Sun 1982; Hart & Zindler 1986; Wänke 1981) and of the continental crust (Taylor & McLennan 1985) tend to reaffirm the more stringent limit of $m_d \leqslant 0.5$. It is noteworthy that this estimate agrees well with values derived from the enrichment of the highly incompatible elements (Rb, Ba, U, Th, etc.) in the continental crust, if it is assumed that these elements have been stripped nearly quantitatively from the depleted portion of the mantle.

A very important limitation of this mass balance is the assumption that the continental crust is the only enriched reservoir. This is known to be incorrect, because the OIBS are also derived from enriched reservoirs of unknown mass, and these were necessarily neglected in the above mass balance. Indeed, some authors have inferred that the OIB sources constitute a large fraction of the mantle (Weaver 1985). If this were so, than there might not be any remaining primitive reservoir, the entire mantle would be differentiated (either depleted or enriched) and this might eliminate the more-popular two-layer mantle models.

Hofmann *et al.* (1986) found that the concentration ratios Nb/U and Ce/Pb could be used for independent mass balances, which explicitly include the OIB source reservoirs. The advantage of these ratios over all the isotopic tracers is that they do *not* discriminate between MORB and

OIB sources. Instead, they discriminate between the *combined* MORB plus OIB source, the continental crust, and the primitive mantle. Figure 2 shows the resulting mass fraction of the combined MORB plus OIB source reservoir, characterized by Nb/U = 47 and Ce/Pb = 25, as a function of the concentrations of U and Pb in the continental crust. Also shown are two estimates of crustal concentrations of U and Pb by Taylor & McLennan (1985, 'T.M.') and Zindler & Hart (1986, 'Z.H.'). The mass fractions of residual mantle range from 0.35 to 0.72 and are thus somewhat higher than those obtained by Nd isotopes, but they are still significantly smaller than 1.0. Consequently, the OIB portion of the residual mantle *may* be quantitatively significant, but it is still too small for the combined MORB plus OIB sources to occupy the entire mantle.

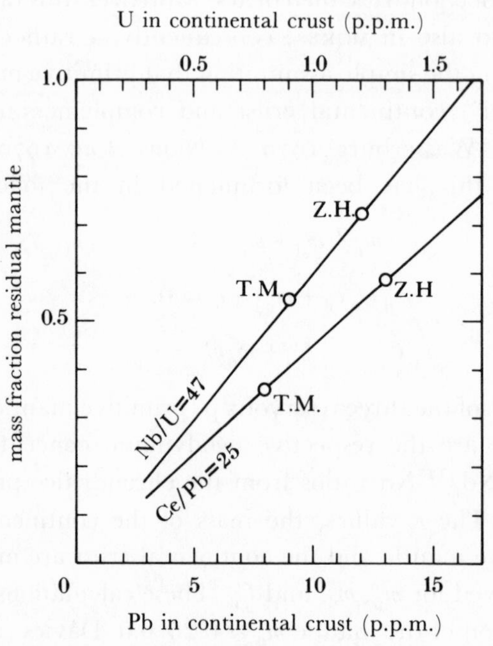

FIGURE 2. Mass fraction of residual mantle (MORB plus OIB source) as a function of U and Pb concentration in the continental crust. The mass balance is calculated using Nb/U = 47 and Ce/Pb = 25 for the residual mantle, Nb/U = 12 and Ce/Pb = 4 for the continental crust, and Nb/U = 30 and Ce/Pb = 9 for the primitive mantle (Hofmann *et al.* 1986). The points labelled T.M. and Z.H. represent concentration estimates of U and Pb in the continental crust by Taylor & McLennan (1985) and Zindler & Hart (1986).

The major remaining uncertainty about the mass balance of crust–mantle reservoirs is that we know relatively little about the composition of the subcontinental lithosphere. Some authors believe that it is also enriched in incompatible elements (see McKenzie & O'Nions 1983; Hawkesworth *et al.* 1984; O'Reilly & Griffin 1988). If its trace-element characteristics are similar to those of the continental crust, even in diluted form, then the effective mass of the continental 'crust', and therefore also the mass fraction of the depleted mantle have been underestimated in all the published mass balances. Partly because of this, the geochemistry of the subcontinental lithosphere is an important object of current investigation.

In summary, it can be concluded that only 30–70 % of residual mantle will balance the chemistry of the continental crust. The remainder of the mantle may still have its primitive composition, unless another enriched reservoir exists that would require more of the mantle to be differentiated. Not enough is known about the subcontinental lithosphere to decide whether it might constitute such a reservoir.

Origin of oib sources and morb source heterogeneities

(a) *Primitive mantle*

Schilling (1973), Hart *et al.* (1973), and Sun & Hanson (1975) recognized the relatively undepleted nature of many OIB sources and, following the ideas of Morgan (1971), suggested that the ocean islands are created by plumes that originate in the deep, undepleted mantle and rise through the depleted upper mantle. The geochemical evidence for this rested mostly on Sr isotopes and rare-earth chemistry of OIBs. These ideas received strong support, first from Nd and later from Hf isotopic data, because these data were well correlated with Sr isotopes and the primitive-mantle value of Nd and Hf isotopic compositions is well constrained by meteorite data. In contrast, the $^{87}Sr/^{86}Sr$ ratio of the primitive mantle cannot be inferred from meteorites, because the Earth has a much lower Rb/Sr (and therefore also $^{87}Sr/^{86}Sr$) ratio than meteorites (Gast 1960). The $^{143}Nd/^{144}Nd$ ratios of OIBs determined by Richard *et al.* (1976), DePaolo & Wasserburg (1976) and O'Nions *et al.* (1977) spanned the range between the depleted MORB values and that of the primitive mantle. Wasserburg & DePaolo (1979) interpreted this using essentially the same plume model formulated earlier by Schilling, Hart, and Sun & Hanson.

Armstrong (1981) disagreed with the above model, pointing out that the isotopic data for Nd and Sr could also be explained by recycled continental crust. This idea was not easily disproved. However, Pb isotopes of both old and young continent-derived sediments tend to be higher in (relative) ^{207}Pb abundance than most (although not all) OIBs (see White *et al.* 1985; Sun 1980). On the other hand, Pb isotopes are even less consistent with OIBs being derived from a primitive reservoir. This may be seen in figure 3, which shows the Pb-isotopic data of continental rocks compiled by Zartman & Doe (1981), compared with data for MORBs from Ito *et al.* (1987) and OIB data from the compilation of Zindler & Hart (1986). It is obvious that continental Pb isotopes follow a separate and significantly steeper trend than oceanic Pb isotopes, and most OIBs are too low in ^{207}Pb to be derived from continental material. Moreover,

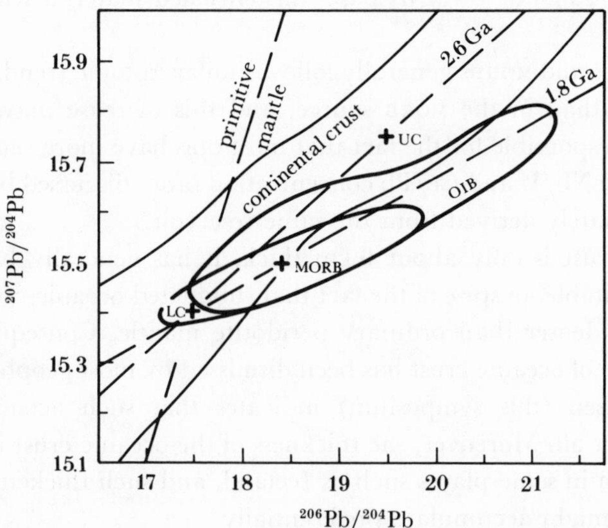

FIGURE 3. Pb-isotopic compositions and model ages of the major crustal reservoirs. The estimated average compositions of the upper (UC) and lower (LC) continental crust according to Zartman & Doe (1981) and MORB according to the new compilation by Ito *et al.* (1987) are shown by crosses. The OIB data are from the compilation of Zindler & Hart (1986). The bulk composition of any closed U–Pb system formed 4.57 Ga ago must lie on the line labelled primitive mantle.

with very few exceptions, the oceanic Pb isotopes are also far too radiogenic to be derived from an undifferentiated U–Pb reservoir, which is defined by the line conventionally called the *geochron*. The only way that these results could be reconciled with a 'primitive' mantle source for OIBS has been to postulate that this deep-mantle reservoir has lost lead to the core several hundred million years *after* the accretion of the Earth (a process termed 'core pumping' by Allègre *et al.* 1982). Core pumping was finally laid to rest by Newsom *et al.* (1986) who showed that OIBS have the same ratios of Pb and Mo to light rare-earth elements as MORBS. If Pb had moved to the core from OIB sources but not from MORB sources, then the strongly siderophile element Mo should have been removed even more efficiently, and the concentration ratios of both elements with the distinctly non-siderophile rare-earth elements should be lower in OIBS than in MORBS. Contrary to these expectations, Pb/Ce and Mo/Pr ratios were found to be identical in MORBS and in OIBS, and therefore removal of lead to the core cannot account for the differences in U/Pb ratios between MORB and OIB sources.

The dilemmas presented by the Pb isotopes were reinforced by Nd-isotopic compositions of some OIBS that do not lie on possible mixing lines between primitive and depleted mantle. Nd-isotope data from Kerguelen and Tristan da Cunha require sources that are enriched in light rare-earth elements relative to a primitive-mantle source.

(*b*) *Recycling*

The increasing difficulties to reconcile the simple models with the geochemical data prompted Hofmann & White (1980, 1982), Chase (1981) and Ringwood (1982) to suggest a recycling model that differs from that of Armstrong in that it emphasizes oceanic crust and lithosphere rather than the continental crust. The oceanic crust is cumulatively by far the largest volume of enriched material. At current production rates, it takes only about 300 Ma to make a volume of oceanic crust equivalent to that of the entire continental crust. If even a small portion of this oceanic crust is stored in the mantle without being completely rehomogenized with ordinary mantle, and if the preferred place of storage is in a thermal boundary layer within the mantle or at its base, this enriched material will make a suitable source for most OIBS.

Figure 3 shows that OIBS and MORBS generally follow similar isotopic trends. U/Pb ratios are generally higher in OIBS than in the MORB source, and this increase may ultimately (after 1–2 Ga, on average) be responsible for the fact that most OIBS have more radiogenic lead than most MORBS. Similarly, the Nb/U and Ce/Pb concentration ratios discussed below indicate that OIBS and MORBS are ultimately derived from the same reservoir.

Because the oceanic crust is only about 6 km thick, it has generally been believed that rehomogenization is inevitable, in spite of the fact that subducted oceanic crust is converted to eclogite and is therefore denser than ordinary peridotite mantle. Consequently the idea of accumulation and storage of oceanic crust has been dismissed by most geophysicists. However, the analysis of Christensen (this symposium) indicates that such accumulation may be mechanically feasible after all. Moreover, the thickness of the oceanic crust is not uniform but reaches as much as 20 km in some places such as Iceland, and such thickened pieces of dense subducted oceanic crust might accumulate preferentially.

Yet another recycling model was offered by McKenzie & O'Nions (1983), who proposed recycling of the subcontinental lithosphere. The critical assumption for this model is that this lithosphere has an appropriately (trace-element) enriched composition. There is indeed evidence for enrichment found in many mantle xenoliths, but there is also at least some evidence that the type of enrichment predominating in the continental lithosphere shows the deficiency in Ti, Nb and Ta, which is characteristic of the continental crust itself (Hawkesworth *et al.* 1987). If this type of enrichment characterized the entire subcontinental lithosphere, it would not be a suitable source of OIBS for the same reasons given above for the continental crust. Still, not enough is known about the subcontinental lithosphere to make a definitive judgement.

(*c*) *Metasomatism*

Other authors proposed that OIB sources are created by ancient or recent metasomatism in the mantle (Menzies & Murthy 1980; Wass & Rogers 1980; Bailey 1982; Vollmer 1983). There are reasons to believe that metasomatic transport in the mantle is restricted to the mesoscale (as defined above) or less. Extensive, large-scale metasomatism would almost certainly create large-scale zoning and extreme chemical heterogeneities through chromatographic mechanisms (Hofmann 1986), whereas real oceanic basalts are characterized by variable enrichment or depletion but otherwise highly coherent trace-element ratios. In particular, some elements of rather different chemical affinities form nearly uniform ratios in all oceanic basalts. Examples of these are Ba/Rb, Zr/Sm, Sr/Ce, P/Nd, K/U, Nb/U and Ce/Pb (Hofmann & White 1983; Hofmann *et al.* 1986; Jochum *et al.* 1983, 1988). It is very difficult to believe that these element pairs would not be separated during any large-scale metasomatic transport. On the other hand, metasomatism certainly does occur in the mantle, and its effects have been studied extensively in mantle xenoliths. If this occurs locally or on the mesoscale, such as in a subduction environment, it may ultimately affect the composition of significant parts of the subcontinental lithosphere. Partial melting may then 'take advantage' of the general enrichment, but average out the chemical heterogeneities created by the metasomatic process.

Additional geochemical evidence

(*a*) *Noble-gas isotopes*

The discovery of large quantities of primordial ^{3}He in oceanic basalts (Lupton & Craig 1975) demonstrated that the Earth has never been completely outgassed, even if its silicate mantle formed an early, extensive magma ocean. The ^{3}He is not recycled from the atmosphere into the mantle, because He entering the atmosphere from the mantle escapes rapidly into space. Although ^{3}He is also produced in the solid Earth spallogenically and by neutron reactions, the current consensus is that the bulk of the ^{3}He emanating from the mantle is primordial. The ^{3}He/^{4}He ratios in MORBS are 10, in OIBS up to 30 times higher than in the current atmosphere. This has revived the ideas that OIBS may be derived from a primitive mantle reservoir after all (Kurz *et al.* 1982; Staudacher *et al.* 1986). In particular, these data appeared to provide powerful arguments against all recycling models. Subsequently, it was recognized that He isotopes correlate with ^{87}Sr/^{86}Sr in some regions but not in others. This led Kurz *et al.* (1982) to suggest that some OIBS originate from a primitive-mantle reservoir, whereas others come from recycled crust, and this view is still popular at present. Recently, high ^{3}He/^{4}He ratios have also been found in crustal xenoliths (Porcelli *et al.* 1986), and this indicated that helium may migrate through the mantle. Thus, the significance of ^{3}He with regard to any of the current models is uncertain at best. The same can be said for Ar, Ne and

Xe isotopes. For example, some mantle samples contain excess ^{129}Xe, which must have been inherited from the first 10^8 years of Earth history, but the significance of this is not yet well understood.

(*b*) *Nb and Pb*

These elements have the unusual property that they distinguish sharply between processes forming continental crust and those forming oceanic crust. By searching for other incompatible elements that would best correlate with these two elements, it was found that Nb/U ratios are the same in MORBS and in OIBS, and the same is true for Pb/Ce ratios (Hofmann *et al.* 1986; Newsom *et al.* 1986). Such uniform ratios may be regarded as 'quasi-isotopic' as explained earlier, that is they are identical in the basalt and its source. This in itself is not surprising, because about a dozen such element pairs are known (Jochum *et al.* 1988). What is surprising is the fact that these two concentration ratios are *not* the same as the corresponding ratios of the primitive mantle, as is observed for other uniform ratios such as Zr/Hf, Nb/Ta, Y/Ho and others. The value of $Nb/U = 47 \pm 10$ in oceanic basalts (exclusive of subduction related volcanics) is significantly higher than the primitive mantle value of $Nb/U = 30$ and the value of the continental crust of $Nb/U \approx 12$. Many island arc volcanics have even lower values. Very similar relations hold for Ce/Pb ratios. The significance of this to the crust–mantle mass balance has already been discussed. Of equal importance is the constraint these data provide for the origin of the OIB sources. It means that the OIB sources have undergone the same differentiation process as the MORB sources, so that *both* are now complementary to the continental crust. On the basis of this constraint, a primitive mantle source can be ruled out even in those cases, such as certain Hawaiian basalts (such as the Koolau volcanics) that are distinguished by high ^3He and apparently nearly primitive Sr, Nd, and Pb isotopic compositions. Similarly, a source dominated by recycled continental crust can also be ruled out, because such material would cause the Nb/U and Ce/Pb to be lower, not higher than primitive. This does not mean that the OIB sources must be totally devoid of primitive or continental components, it only means that such components must be minor ones at most. Weaver *et al.* (1987) have shown that the islands of Gough and Tristan da Cunha have somewhat lower Nb/Th and Nb/La ratios than other islands and proposed that their sources are mainly subducted oceanic crust contaminated with a few percent of recycled pelagic sediment. Such a sedimentary component is expected to be introduced into the mantle during subduction, and the only surprise is that its geochemical signature is not more commonly found in OIBS.

Residence time of lead in the MORB source

Galer & O'Nions (1985) have recently calculated the age of the depletion of Th relative to U in the MORB source mantle from the relative abundances of ^{208}Pb and ^{206}Pb in MORB, assuming a simplified, single-event mantle evolution. If the ratio of their respective parent nuclides, $^{232}Th/^{238}U$ changed from an original (primitive mantle) value of 3.9 to 2.5 at the time of the depletion event, then this event took place as recently as 600 ± 200 Ga ago, in the present-day MORB source. This result is contrary to expectations derived from relations of $^{207}Pb/^{204}Pb$ against $^{206}Pb/^{204}Pb$ (figure 3), which indicate a differentiation age of 1.8 Ga for MORB plus OIB, or from the mean age of the continental crust, which is about 2.6 Ga by using crustal Pb isotopes (figure 3), or at least 1.5 Ga by using the crust–mantle evolution models of Jacobson & Wasserburg (1979). To reconcile these observations, Galer & O'Nions formulated a new model of mantle differentiation, in which the upper (MORB source) mantle is in a steady state and therefore contains no information on specific differentiation events. Fresh, primitive mantle material is added to this steady-state reservoir by entrainment from the lower mantle, and this is balanced by losses to the continental crust. The steady-state concentration of each element is determined only by its bulk partitioning between the upper mantle and the continental crust.

This model is still quite new, and several aspects remain to be evaluated. For example, Allègre *et al.* (1986) have estimated a terrestrial Th/U of 4.2 rather than 3.9. If correct, the higher ratio would increase the calculated differentiation time from 600 Ma to about 1200 Ma, and this would be in reasonable agreement with mean differentiation ages calculated from other isotopic systems. Another open question concerns the evidence for rapid mixing in the upper mantle, which is required if a steady state is to be attained.

State of mixing in the mantle

It is often argued that the upper mantle is a well-homogenized reservoir, and this assertion is supported by theoretical arguments as well as by numerical experiments (see, for example, Hoffman & McKenzie 1985). However, this is not immediately obvious from the geochemical data themselves. Taken at face value, the isotopic and trace-element data for all oceanic basalts indicate a highly heterogenous, poorly mixed mantle. For example, $^{206}Pb/^{204}Pb$ ratios range from 17 to 21, and this range is larger than the difference between Zartman & Doe's (1981) best estimates of the upper and lower crustal averages in continental rocks (figure 3). The apparent state of homogeneity improves only moderately if the sampling is restricted to the so-called N-type MORB, that is, if all OIBs and suspected 'plume-type' MORBs are removed from the sample considered. The N-type MORB samples analysed and reviewed by Ito *et al.* (1987) have a range of $^{206}Pb/^{204}Pb = 17$ to 19.4. Similarly, these same samples have Ba concentrations between 1.5 p.p.m. and 32 p.p.m. (by mass) (Jochum & Hofmann 1989). This range exceeds considerably that expected from differences in the degree of melting and fractional crystallization and most likely reflects differences in source composition. Finally, extreme isotopic compositions have been observed such as $^{40}Ar/^{36}Ar = 25000$ (Staudacher *et al.* 1986) in some depleted MORBs. Entrainment and rapid homogenization of even small amounts of mantle material with a primitive or atmospheric ratio of $^{40}Ar/^{36}Ar \approx 300$ would not allow such high ratios to survive, unless all the entrained Ar were vented immediately to the atmosphere. Thus, the idea of a well-mixed reservoir cannot be deduced directly from the data; rather it must be reconciled with the data by assuming that the observed heterogeneities are not stored within the mixing bowl but are introduced from the outside and only a short time before they reappear in the various oceanic basalts.

Conclusions

The heterogeneities that can be detected and identified by geochemical means are not directly detectable by seismic techniques. Future progress in connecting the two types of evidence will therefore depend on establishing correlations between seismic velocity structure and geochemistry.

The salient conclusions from geochemical evidence are as follows.

(1) Less than 70%, perhaps as little as 25%, of the mantle mass is sufficient to balance the chemical composition of the continental crust, but the remaining primitive (or less depleted) mantle reservoir does not appear to be sampled directly by plumes or by MORBS.

(2) OIB sources are most easily explained by some type of recycling. However, typical continental crust or continent-derived sediments are either absent or only a minor component of OIB sources. This reduces the choice of recycling candidates to delaminated subcontinental lithosphere and/or subducted oceanic crust and lithosphere.

(3) Metasomatism may have modified the composition of the OIB sources but is judged to be a local phenomenon, which is probably important mainly during subduction and the concomitant devolatilization of the subducted crust. In this way, it may have affected the composition of the subcontinental lithosphere.

(4) The bulk of the continental crust was separated from the mantle before the residual mantle was differentiated into MORB and OIB sources, but the residence times of heterogeneities in this residual mantle remain uncertain.

(5) The popular notion that the MORB source is a well-mixed reservoir can be reconciled with geochemical data only if one assumes that all the heterogeneities are stored outside that reservoir. The geochemical peculiarities common to both MORBS and OIBS (Pb-isotope trends, Nb/U and Ce/Pb ratos) favour the interpretation that both the depleted and the enriched parts of the mantle belong to a common mantle reservoir, which was once homogenized but has since become quite heterogeneous through secondary differentiation processes.

I thank Keith O'Nions, F.R.S., for reviewing the first version of this manuscript and for giving me *carte blanche* to say what I want. I especially thank Bill McDonough and Steve Goldstein for letting themselves be pressured into last-minute reviews of the revised manuscript and for suggesting significant improvements.

References

Allègre, C. J. & Condomines, M. 1982 *Nature, Lond.* **299**, 21–29.
Allègre, C. J., Dupré, B. & Brévart, O. 1982 *Phil. Trans. R. Soc. Lond.* A **306**, 49–59.
Allègre, C. J., Dupré, B. & Lewin, E. 1986 *Chem. Geol.* **56**, 219–227.
Allègre, C. J., Hart, S. R. & Minster, J. E. 1983 *Earth planet, Sci. Lett.* **66**, 191–213.
Allègre, C. J. & Turcotte, D. L. 1986 *Nature, Lond.* **323**, 123–127.
Armstrong, R. L. 1981 *Phil. Trans. R. Soc. Lond.* A **301**, 443–472.
Bailey, D. K. 1982 *Nature, Lond.* **296**, 525–530.
Batiza, R. & Vanko, R. 1984 *J. geophys. Res.* **89**, 11235–11260.
Chase, C. G. 1981 *Earth planet. Sci. Lett.* **52**, 277–284.
Chen, C. Y. & Frey, F. A. 1985 *J. geophys. Res.* **90**, 8743–8768.
Davies, G. F. 1981 *Nature, Lond.* **290**, 208–213.
DePaolo, D. J. 1980 *Geochim. cosmochim. Acta* **44**, 1185–1196.
DePaolo, D. J. & Wasserburg, G. J. 1976 *Geophys. Res. Lett.* **3**, 743–746.
Feigenson, M. D. 1984 *Contr. Miner. Petr.* **87**, 109–119.
Fitton, J. G. & Dunlop, H. M. 1985 *Earth. planet. Sci. Lett.* **72**, 23–38.
Galer, S. J. G. & O'Nions, R. K. 1985 *Nature, Lond.* **316**, 778–782.
Galer, S. J. G. & O'Nions, R. K. 1986 *Chem. Geol.* **56**, 45–61.
Gast, P. W. 1960 *J. geophys. Res.* **65**, 1287–1297.
Hanson, G. N. 1977 *J. geol. Soc. Lond.* **134**, 235–253.
Hart, S. R., Schilling, J.-G. & Powell, J. L. 1973 *Nature, Lond.* **246**, 104–107.
Hart, S. R. & Zindler, A. 1986 *Chem. Geol.* **57**, 247–267.
Hart, S. R. & Zindler, A. 1988 In *Mantle convection* (ed. D. Peltier). London: Gordon Breach.
Hawkesworth, C. J., Rogers, N. W., van Calsteren, P. W. C. & Menzies, M. A. 1984 *Nature, Lond.* **311**, 331–335.

Hawkesworth, C. J., van Calsteren, P., Rogers, N. W. & Menzies, M. A. 1987 In *Mantle metasomatism* (ed. M. A. Menzies & C. J. Hawkesworth), pp. 365–388. London: Academic Press.
Hoffman, N. R. A. & McKenzie, D. P. 1985 *Geophys. Jl R. astr. Soc.* **82**, 163–206.
Hofmann, A. W. 1986 *Chem. Geol.* **57**, 17–30.
Hofmann, A. W. 1988 *Earth. planet. Sci. Lett. Crafoord Special Issue.* (In the press.)
Hofmann, A. W., Feigenson, M. D. & Raczek, I. 1987 *Contr. Miner. Petr.* **95**, 114–122.
Hofmann, A. W. & Hart, S. R. 1978 *Earth. planet. Sci. Lett.* **38**, 44–62.
Hofmann, A. W., Jochum, K. P., Seufert, M. & White, W. M. 1986 *Earth. planet. Sci. Lett.* **79**, 33–45.
Hofmann, A. W. & White, W. M. 1980 *Carnegie Instn Wash. Yb.* **79**, 477–483.
Hofmann, A. W. & White, W. M. 1982 *Earth. planet. Sci. Lett.* **57**, 421–436.
Hofmann, A. W. & White, W. M. 1983 *Z. Naturf.* a**38**, 256–266.
Ito, E., White, W. M. & Göpel, C. 1987 *Chem. Geol.* **62**, 157–176.
Jacobsen, S. B. & Wasserburg, G. J. 1979 *J. geophys. Res.* **84**, 7411–7427.
Jochum, K.-P., Hofmann, A. W., Ito, E., Seufert, H. M. & White, W. M. 1983 *Nature, Lond.* **306**, 431–436.
Jochum, K.-P. & Hofmann, A. W. 1989 *The chemistry of oceanic basalts.* (In preparation.)
Klein, E. M. & Langmuir, C. H. 1987 *J. geophys. Res.* **92**, 8089–8115.
Kurz, M. D., Jenkins, W. J. & Hart, S. R. 1982 *Nature, Lond.* **297**, 43–47.
Lanphere, M. A., Dalrymple, G. B. & Clague, D. A. 1980 *Initial Reports Deep Sea Drilling Project*, vol. 55, 695–706.
Lupton, J. E. & Craig, H. 1975 *Earth. planet. Sci. Lett.* **26**, 133–139.
McKenzie, D. & O'Nions, R. K. 1983 *Nature, Lond.* **301**, 229–231.
Menzies, M. A. & Murthy, V. R. 1980 *Am. J. Sci.* A**280**, 622–638.
Morgan, W. J. 1971 *Nature, Lond.* **230**, 42–43.
Newsom, H. E., White, W. M., Jochum, K.-P. & Hofmann, A. W. 1986 *Earth. planet. Sci. Lett.* **80**, 299–313.
O'Hara, M. J. & Mathews, R. E. 1981 *J. geol. Soc. Lond.* **138**, 237–277.
O'Nions, R. K., Evensen, N. M. & Hamilton, P. J. 1979 *J. geophys. Res.* **84**, 6091–6101.
O'Nions, R. K., Hamilton, P. J. & Evensen, N. M. 1977 *Earth. planet. Sci. Lett.* **34**, 13–22.
O'Reilly, S. Y. & Griffin, W. L. 1988 *Geochim. cosmochim. Acta* **52**, 433–447.
Patchett, P. J. & Tatsumoto, M. 1980 *Geophys. Res. Lett.* **7**, 1077–1080.
Porcelli, D. R., O'Nions, R. K. & O'Reilly, S. Y. 1986 *Chem. Geol.* **54**, 237–249.
Reisberg, L. & Zindler, A. 1986 *Earth. planet. Sci. Lett.* **81**, 29–45.
Richard, P., Schimizu, N. & Allègre, C. J. 1976 *Earth. planet. Sci. Lett.* **31**, 269–278.
Ringwood, A. E. 1982 *J. Geol.* **90**, 611–643.
Schilling, J.-G. 1973 *Nature, Lond.* **242**, 565–571.
Schilling, J.-G., Zajac, M., Evans, R., Johnston, T., White, W., Devine, J. D. & Kingsley, R. 1983 *Am. J. Sci.* **283**, 510–586.
Sleep, N. H. 1984 *J. geophys. Res.* **12**, 10029–10041.
Staudacher, T., Kurz, M. D. & Allègre, C. J. 1986 *Chem. Geol.* **56**, 193–205.
Sun, S.-S. 1980 *Phil. Trans. R. Soc. Lond.* A**297**, 409–445.
Sun, S.-S. 1982 *Geochim. cosmochim. Acta* **46**, 179–192.
Sun, S.-S. & Hanson, G. N. 1975 *Geology* **3**, 297–302.
Tatsumoto, M., Hedge, C. E. & Engel, A. E. J. 1965 *Science, Wash.* **150**, 886–888.
Taylor, S. R. & McLennan, S. M. 1985 *The continental crust: its composition and evolution.* Blackwell Scientific Publ. (312 pages.)
Vollmer, R. 1983 *Geology* **11**, 452–454.
Waff, H. S. & Bulau, J. R. 1979 *J. geophys. Res.* **84**, 6109–6114.
Wänke, H. 1981 *Phil. Trans. R. Soc. Lond.* A**303**, 287–302.
Wass, S. Y. & Rogers, N. W. 1980 *Geochim. cosmochim. Acta* **44**, 1811–1824.
Wasserburg, G. J. & DePaolo, D. J. 1979 *Proc. natn. Acad. Sci. U.S.A.* **76**, 3594–3598.
Watson, E. B., BenOtham, D., Luck, J. M. & Hofmann, A. W. 1987 *Chem. Geol.* **62**, 191–208.
Weaver, B. L. 1985 *Eos, Wash.* **66**, 1113 (abstract).
Weaver, B. L., Wood, D. A., Tarney, J. & Joron, J. L. 1987 In *Alkaline igneous rocks* (ed. J. G. Fitton & B. G. J. Upton) Geol. Soc. Spec. Publ. 30, pp. 253–267.
White, W. M. 1985 *Geology* **13**, 115–118.
White, W. M., Dupré, B. & Vidal, P. 1985 *Geochim. cosmochim. Acta* **49**, 1875–1886.
White, W. M. & Hofmann, A. W. 1982 *Nature, Lond.* **296**, 821–825.
Zartman, R. E. & Doe, B. R. 1981 *Tectonophysics* **75**, 135–162.
Zindler, A. & Hart, S. R. 1986 *A. Rev. Earth. planet. Sci.* **14**, 493–571.
Zindler, A., Staudigel, H. & Batiza, R. 1984 *Earth. planet. Sci. Lett.* **70**, 175–195.

Phil. Trans. R. Soc. Lond. A **328**, 441–442 (1989)
Printed in Great Britain

Chemical boundary layers of the mantle and core [abstract only]

By T. H. Jordan

Department of Earth, Atmospheric, and Planetary Sciences, Massachusetts Institute of Technology, Cambridge, Massachusetts 02139, *U.S.A.*

There is now sufficient information from seismological mapping of the Earth's deep interior to draw some preliminary conclusions regarding nature of large-scale mantle flow. This paper examines three features of mantle heterogeneity.

Seismological studies confirm the existence of a thick (more than 300 km) thermal boundary layer (TBL) beneath the ancient cratonic nuclei. Petrological and gravimetric data imply that the continental TBL is stabilized against convective disruption by a buoyant, viscous, chemical boundary layer (CBL) depleted in Fe and Al relative to Mg. Geothermal constraints require high heat production within the CBL and low heat flow through its base, indicating that the CBL has been recharged by large-ion lithophile (LIL) elements after primary depletion events. Formation of this continental tectosphere cannot be simple conductive cooling, as in the oceans, but must involve several stages characterized by different timescales, terminating with crustal stabilization; the advective thickening of a basalt-depleted, LIL-rich CBL in episodes of compressive orogenesis (e.g. supercontinent assembly) may be an important mechanism for tectospheric consolidation. The stability and low basal heat flow of the cratonic CBL are evidence that the positions of the continents through time are coupled to the upward flow of material from the deep mantle.

In the transition zone and lower mantle, the aspherical variations in seismic velocities are dominated by the downward and outward flow of cold material beneath areas of active subduction. Residual-sphere analysis of P-wave and S-wave data from deep-focus earthquakes indicates the presence of anomalous high-velocity material extending into the lower mantle below the seismically active portion of the slab. The dimensions and orientations of these high-velocity zones are consistent with the hypothesis that essentially all slab material is sinking through the 650 km discontinuity into the lower mantle; this flux is sufficient to recycle the entire volume of the upper mantle into the lower mantle in less than 10^9 years and cannot be reconciled with chemical–convective stratification at this boundary. Global tomographic models and regional studies delineate high-velocity anomalies in the lower mantle beneath Cainozoic subduction zones consistent with the constraints established by the residual-sphere analysis.

Large-amplitude, low-wavenumber heterogeneity has recently been discovered in the vicinity of the core–mantle interface. As in the case of the free surface, it is difficult to account for the magnitude of the asphericity without postulating the existence of a laterally heterogeneous CBL on one or both sides of the core–mantle interface. The processes that potentially contribute to CBL formation include the accumulation of 'dregs' at the base of the mantle (e.g. subducted oceanic crust), the accumulation of 'slag' at the top of the outer core (e.g. from inner-core differentiation), chemical reactions at the interface, and primary chemical layering inherited from inhomogeneous accretion and/or core formation. This system of CBLs may interact strongly with mantle convection and play a crucial role in coupling it to convection in the core.

In summary, the low-wavenumber aspherical heterogeneity of the mantle and outermost core is dominated by boundary layers formed at the free surface and core–mantle interface, and chemical heterogeneity appears to play an important role in governing the configuration of these boundary layers. The seismological data appear to be consistent with an Earth comprising four major dynamical systems: the two convecting shells of the mantle and core, and the two CBLs at the free surface and core–mantle interface. Strong interactions among the low-wavenumber states of these four systems offer new possibilities for explaining the Earth's large-scale, long-term behaviour.